# Sun,
# Sand,
# and
# Sea Serpents

David Goudsward

Anomalist Books
*San Antonio * Charlottesville*

An Original Publication of ANOMALIST BOOKS
**Sun, Sand, and Sea Serpents**
Copyright 2020 by David Goudsward
ISBN: 978-1-949501-11-7

On the cover: A sea monster prowls the Gulf of Mexico in a detail from the 1565 "Le Moyne Map" of Florida and Cuba, drawn in 1563 by French artist Jacques Le Moyne de Morgues and first published by Theodore de Bry in 1591.

An earlier version of Chapter 6 appeared in the *International Cryptozoology Society Journal* 1 (2017).

Book Design: Seale Studios

For information about the publisher, go to AnomalistBooks.com, or write to: Anomalist Books, 5150 Broadway #108, San Antonio, TX 78209

# Contents

For Tim
VPUSMMW
HHS78/79

# Foreword

Florida is a world of its own, far different from everywhere else in the United States, and definitely not in the slightest like southern California nor Hawaii, which may be mildly similar in yearly temperatures but not in mood. The breezes, reptiles, and people of Florida daily let you know you are in a different existence. And so do the cryptids.

I feel responsible, at least partially, for stimulating the initial creation of this book. During and after the First International Cryptozoology Museum Conference in Saint Augustine, Florida, in January 2016, I asked David Goudsward if he would contribute to our proceedings with a paper on the events surrounding the area of the beaching of the Giant Octopus of 1896.

I had been to St. Augustine several times, and on my 2011 trip I had gone to every tourist shop to ask if they had any postcards, books, or souvenirs related to what is celebrated in cryptozoology as a major event—"St. Augustine Giant Octopus." While popular souvenir shops are full of trinkets, tee-shirts, and trivial merchandise on pirates, mermaids, and the Spanish discovery and occupation of the area, the St. Augustine shop owners had nothing artistic or tourist-related to the 1896 Giant Octopus. They were totally unaware of the incident itself. Some people I interviewed even thought I was making up the story.

The 2016 conference was an effort to heighten awareness of that event, and my request to Goudsward to look into the beaching was part of that effort. Now you have to understand, I knew what I was getting myself into, since the writer, researcher, and investigator who is Dave Goudsward does not take such challenges lightly. As opposed to merely retelling stories, or rewriting others' studies of specific cases, he is one of the best researchers I know for digging up original sources. I'm glad to see that Goudsward's original Giant Octopus article has morphed, in this book, into a broader in-depth investigation of the sea serpents, cryptids, and out-of-place fauna of Florida, the Caribbean, and the southeast. He has cast a wide net to find various nuggets of truth about many cryptid mysteries.

One of the best ways to know an author has done his job is to put a microscope on one instance of the story that is very personal. In the early 2000s, I conducted a mini-expedition in a rugged old boat on the St. Johns River in Florida, interviewing locals and actively looking for Pinky, the "living dinosaur" of these parts. I, along with a number of other cryptozoologists, have mentioned in various books and articles the unique river monsters seen up and down the St. Johns River in Florida, known by the collective name as "Pinky," since the 1970s.

The nickname came about when on May 10, 1975, Charles and Dorothy Abram were fishing with three friends when they saw the head and neck of a large pink animal about 20 feet away. It turned its head to look at them and then submerged after about eight seconds. Eyewitness Dorothy Abram said it looked like "a dinosaur with its skin pulled back so all the bones were showing…pink, sort of the color of boiled shrimp."

Their description was concise and yet remarkable. They said it had a pink color, a human-sized head, small horns with knobby tips, bones showed that through the skin, flappy skin on the sides of the head, a three-feet-long neck, and slanted, dark eyes.

Most reports of Pinky are dated to the 1950s. But the Native peoples of Florida have told about Pinky and other related sightings of river monsters as long ago as 1888.

Sometimes when investigating cryptids, it's not how you ask but who and what you are asking about that gets more insightful answers. In 2008, from talking to the late Charlie Carlson, a colorful and friendly character, I gained a deeper insight. "Pinky," which has been used in the cryptozoological literature since 1975, is not used in Florida. Charlie gave me one of the "lightbulb" moments you get in life.

In this area of Florida, where I have been concentrating my on-the-water and shore searching, from the Blue Springs State Park (highly recommended) to Lake Monroe and beyond, I found this cryptid is not called "Pinky" but the "St. Johns Monster." Hiding in plain sight, I feel, sightings have been happening up and down this area of the St. Johns River. Charlie shared that he has not been writing about these encounters as "Pinky," per se, but merely as "lake monster" or "river monster" reports. So too would Dave.

In the search for cryptids, various lines of inquiry must be pursued to discover whether possibly mundane observations may have been transformed into the extraordinary. Goudsward's discovery of mistakes and mix-ups in this realm has helped to clarify the record, and for that I am thankful.

All told, Dave Goudsward's extended examination of all the cryptids of Florida and the surrounding area is enlightening. Enjoy the quest!

Loren Coleman
Sebastian, Florida
February 14, 2020

# Introduction

The problem with sea serpents is they have no sense of geography. A sea serpent spotted off Fort Lauderdale sees no difference between Floridian and Caribbean waters. So, although this book's focus may appear Floridian, it is only because Florida has a better publicity team. Sea Serpents are regional beasts, and the serpents spotted the mouth of the St. Johns River may be the same cryptid as Georgia's Altamaha-ha and various Caribbean creatures. Although zoologist Bernard Heuvelmans believed sea serpents of the warm climes differ from those of the northern cold waters, there appear to be reports of similar 50-to-100 foot serpent sightings along the coast that creates an unbroken line from South America into the Caribbean and all the way up into the Canadian Maritimes. This book covers southeastern North America in the broadest sense of the term. Should there be a second volume, it would focus on New England but include the Canadian coast and the mid-Atlantic region where it would dovetail with the region covered in this book.

Sea monsters and I have a long history. You see, back in the early and mid-1960s, when I was a mere lad, my Saturday morning cartoon of choice was *The Beany And Cecil Show*. The 78 endlessly repeated cartoon adventures of Beany Boy and his buddy, Cecil the Sea-Sick Sea Serpent, were eagerly engrained in my pre-adolescent brain. And yes, I had the Cecil doll, complete with a pull-string voicebox, googly eyes, and chartreuse-dyed plush of undetermined toxicity. In fact, my mother still blames creator Bob Clampett for my love of slapstick and bad puns, but that's another story.

When I got older, if not wiser, my first vacation as a tax-paying, gainfully employed adult was to Loch Ness. It was inevitable. While still in high school, Robert Rines had taken "the flipper photograph" at Loch Ness, and I had met Marty Klein, "The Father of Side Scan Sonar," whose labs were less than ten miles from the high school where I had eagerly devoured all reports coming in from the Highlands. What was less inevitable, and probably avoidable, was tripping on a rain-slick rock and falling into Loch Ness. Fun fact: The waters of Loch Ness are 42°F year-round. This is probably how I ended up living in South Florida where water is never below 70°F and when the forecast calls for 50°F temperatures, they start opening the emergency shelters.

More than once, I've had the passing thought that the Gloucester Sea Serpent didn't vanish, it moved in with southern relatives for similar Fahrenheit considerations as mine. Florida has always been a hotbed for sea monsters, but there is a problem with the term "Florida Sea Monster" itself. In the early days of tourism, before air conditioning and giant mice, hyperbole was king. You didn't catch a large fish, you landed a sea monster. And if a seaside hotel had guests who saw a sea serpent, it "somehow" made the newspapers. It became a joke in the Florida newspapers that they knew when tourist season had started by the first report of a sea serpent. Even today, stories of sea monsters continue to come in, from tourists and recent transplants, of mostly misidentified known species. In writing this book, I quickly learned that any interview with a witness that included a variation of the phrase "I know manatees—I've seen them, and I've swum with them, and that was no manatee" is almost certainly going to turn out to be a manatee.

The word *cryptozoology* means the "study of hidden animals." The word was popularized by Heuvelmans as an all-encompassing term to quantify the study of animals that are not formally recognized by academic zoology but could exist by virtue of reports, current and historical, made by untrained local observers. Depending on the personal criteria of the cryptozoologist in question (or his producer's desperation for new scripts), these cryptid creatures could include species presumed to be extinct, or mythological. So, to the casual viewer, cryptids tend to be the sexy, TV-ready beasts such as Bigfoot and Kraken. Somehow, the discovery of a new frog in the Andes isn't good press, yet it is the culmination of the same process as seeking the Loch Ness Monster—exploration, research, and documentation. The ultimate goal of the cryptozoologist is to eliminate a cryptid from the cryptid category, either through debunking reports, correcting misidentifications, or by documenting irrefutable evidence. The latter has seen cryptids turn into recognized species, such as the okapi or the coelacanth.

As with so many things, cryptozoology in the southeast and Caribbean introduces its own complications into the mix. The earliest reports of cryptids predate Florida and the cruise ships of the Caribbean. In his *diario* entry for October 21, 1492, Columbus recorded that he had killed a "serpent" measuring almost five-feet long that he had seen entering a lake on the Bahamian island. Identification of what Columbus killed (and skinned to send the hide back to Ferdinand and Isabella as a gift) has been an ongoing debate. Columbus used other specific terms for snakes and alligators, but "serpent" was more a catch-all phrase, which could be anything from an iguana to a cryptid.

In 1987, Dr. William F. Keegan, the curator of anthropology at the Florida Museum of Natural History, uncovered the femur of the Cuban crocodile (*Crocodylus rhombifer*) from an excavation of the Lucayan Indian settlement on Crooked Island, the same island where Columbus caught his serpent. This not only serves as evidence that the Cuban crocodile survived longer than suspected after the arrival of the Lucayans, it almost certainly identifies the mystery serpent that Columbus encountered.

Further complicating the matter, there is a phenomenon that I, without any affection, refer to as "Bermuda Triangle Syndrome." Back in the late 1970s, Charles Berlitz hit the *New York Times* bestseller list with his first book on the Bermuda Triangle, spawning a veritable tsunami of books on the topic. Early authors would "borrow" and modify Berlitz's own material, which in turn was borrowed and modified by subsequent authors. The culmination of this paranormal version of the Telephone Game occurred when Berlitz published his next book, in which he cited his original material as new disappearances—his content had been modified so drastically that even he couldn't recognize it. That is an increasingly frequent issue in cryptozoology, as social media proliferates incorrect accounts.

For instance, in May 1923, the cable ship *All America* was dismantling cable buoys outside Vera Cruz harbor. Boatswain's Mate Carl Sjostrom was dismantling a cable buoy anchored in 1200-fathom-deep water when a shark struck the buoy and began circling. His wake sent the buoy rolling in all directions and threatening to swamp the small boat where his work team waited. Sjostrom spent two hours clinging to the buoy while the shark brushed the buoy and circled the boat. Sjostrom's team in the boat attempted to chase off the shark with the oars, only to have the shark take toothy bites from them. After two hours, a crewman jammed the splintered oar into the shark's gills, which send the beast off trailing blood. Sjostrom was able to jump back on the boat, and the team hastily returned to the ship—just in time to help battle three giant octopuses that had climbed aboard the *All America*. The story was tremendously popular in the newspapers of the day, with versions telling how Sjostrom barely avoided the snapping jaw of the shark and the dreadful battle against the tentacled horrors. Except that the version recounted here suffers from Bermuda Triangle Syndrome.

Sjostrom was indeed trapped on the buoy while a shark circled the boat, but the rest of the morning was without incident. Later that afternoon, when the *All America* resumed reeling up an old 1881 cable, which was coated with coral, barnacles, and assorted plant debris, it also pulled up three octopuses, which were apparently anchored to the cable. Not giant, malevolent predators, just cephalopods that held on to the wrong thing too long. The cable foreman grabbed a shovel and smacked one, which wrapped two of its arms around the foreman's leg, resulting in knife play and an octopus pummeled to death. The other two were pushed off the side with boathooks. Not a bad story either, but not nearly as newspaper-worthy as a crew battling giant man-eating monsters. Even the 1200 fathom depth is an exaggeration—it was based on Sjostrom's description of where they were anchored; the sea was just over 20 fathoms deep (less than 150 feet deep). That's certainly of little consequence when being stalked by a shark, but it's a depth that would suggest the stowaways that came up with the cable were common octopuses (*Octopus vulgaris*), which have arms up to three-feet long—undoubtedly alarming to find wrapped around your leg, but a far cry from the newspaper version.

Fewer sea monsters are seen today, which could simply be that there are now fewer aquatic curiosities to be unidentified. But fear not, although this book will, with no little regret in some cases, debunk some of the tourist-enticing cryptozoological aquatic marvels, there are still many unknown and unidentified. This is southeast North America, where the weird is as plentiful as the sunshine.

And yes, I still have my Cecil doll but have no plans to revisit Loch Ness.

# Chapter 1
## Living Fossils

Fossils, by definition, are the preserved records of any living thing from a past geological age. With the great antiquity involved in becoming a fossil, the inference is that the creature is now extinct, or has evolved into a current species. Cryptozoologists, however, are eternal optimists—declaring an animal extinct may only mean the creature is harder to locate in the uncharted forests or deep oceans. Such optimism is not always a virtue; blindly accepting every misidentification, second- or third-hand report, indistinct track, or fuzzy photo, diminishes the reliability of legitimate research.

"Living Fossils," species that appear unchanged compared with fossil remains, do exist, ranging from the gingko tree to the red panda, but since they are well-documented, they fall outside the realm of cryptozoology. Cryptozoologists instead look for new creatures to document. This also means that rediscovering recent extinctions, such as the ivory-billed woodpecker, are just as important in cryptozoology as finding Bigfoot or surviving dinosaurs. In Florida, the Everglades may prove to be a haven for living fossils, and one may never know what deep water mystery may be carried with currents onto the beaches. Assuming, of course, there are living fossils left to discover.

### Basking Sharks
The basking shark (*Cetorhinus maximus*) holds a unique place in cryptozoology—it is only mistaken for a sea serpent after its death. The second-largest living shark after the whale shark, it grows to lengths of 40 feet. Harmless filter feeders, the sharks are migratory wanderers in search of plankton. Tiny teeth, enlarged gill slits, and cartilaginous gill rakers are evolutionary developments related to the basking shark's filter-feeding mode of life. But those features also create conditions that Bernard Heuvelmans refers to in his texts as the "pseudo-plesiosaur" effect. The gills of the basking shark rot quickly after death, and when they detach, the shark's jaw falls away as well. This leaves the skull and backbone exposed, which looks like a small head and long neck. Similarly, the shark's spine stops at the upper lobe of the caudal fin (tail), so when the lower

lobe decomposes and falls off, it looks like a long tail. And to fully trigger the pseudo-plesiosaur effect, the shark's pectoral fins (and sometimes the pelvic fins) remain attached, resembling flippers.

In the spring of 1885, Rev. Gustavus E. Gordon of Milwaukee, Wisconsin, was president of the United States Humane Society. The position involved a great deal of travel on behalf of the fledgling organization. On a visit to South Florida, the ship dropped anchor in New River Inlet, an area famous for its prolific fish population that attracted an equally large number of sharks.

While lying at anchor, the anchor became entangled with what appeared to be the carcass of a sea serpent. Its total length was 42 feet, and the girth was six feet. The head was absent; it had two front flippers and a six feet long slender neck. The good reverend immediately realized it must be the remains of an enaliosaurus, an alternative name for a plesiosaur. Gordon had it hauled above high water mark in an attempt to preserve the remains until they could be retrieved. Upon his return to New York, Gordon contacted J. B. Holder, the curator of Marine Zoology of American Museum of Natural History and an expert on whales and Florida sea life. Holder made arrangements for the recovery and transportation of the remains to New York.

Rev. Gordon was convinced he had found proof the plesiosaur was not extinct but was over-optimistic in gauging how safe the shore was, above the high tide mark or not. On August 24, a tropical storm paralleling the coast began strengthening into a hurricane. The remains were swept back out to sea. By the time the museum crew arrived, all that remained were Gordon's rough sketches and his impeccable reputation.

Without physical evidence, all Dr. Holder could do is write up Rev. Gordon's account. It was published in *The Century Magazine*. Further complicating the story, Holder asked the opinion of his associate naturalist Frederic A. Lucas at the Smithsonian. Lucas rendered Gordon's sketches into something decidedly non-plesiosauric—essentially a serpent with vestigial front flippers. Holder decided the drawings were more reminiscent of a clidastes, an extinct genus of marine reptile belonging to the mosasaur family with flipper-like arms and legs and a tail fin. He included a sketch of a recently found clidastes skeleton. Just to further complicate things, the magazine then captioned the Lucas reconstruction as having been found in the wrong river.

The New River Inlet is now gone. The river has been directed into canals that dump into Port Everglades. The location of the outlet, if it still existed, would be in neighboring Hollywood, Florida, roughly where the Hollywood Dog Beach is on North Surf Road. Frederic A. Lucas would find his ardor for reconstructing cryptids somewhat tempered when he became entangled in the giant octopus of St. Augustine debate a few years later (Chapter 6).

As for Rev. Gordon's plesiosaur, it was neither the first nor the last decomposed basking shark to be mistaken for a sea monster. His description matches the description and sketches of other basking shark misidentifications such as the 1808 Stronsay Beast of the Orkney Islands, and a carcass caught by the fishing trawler *Zuiyō Maru* off New Zealand in 1977.

In 2018, a carcass washed up on Wolf Island National Wildlife Refuge near Darien, Georgia, leading to the usual flurry of sea monster claims. It was followed by cries of fraud and theories ranging from another decomposing basking shark to a baby Altamaha-ha, the sea monster in the Altamaha River and adopted mascot of Darien (Chapter 7). It would turn out to be "situational performance art" by a New York-based, self-identified "myth-maker" who specializes in online viral hoaxes. This may be a first in cryptozoology, which is not noted for performance art.

**Frilled Sharks**

In the late 1890s, the panhandle town of Carrabelle was beginning to transform from a deep water fishing port into an industrial center. Sawmills dotted the Carrabelle River and the docks were backlogged with the volume of lumber and turpentine being shipped out. A railroad station carried salted fish to Tallahassee and Georgia, and tourists were using the train to flock to a resort hotel just up the coast.

At 11 a.m. on August 19, 1896, the paddle steamer *Crescent City* was following the shoreline of Dog Island, one of the barrier islands that shelter Apalachicola Bay. She had been booked for a fishing charter with 30 passengers who came by special train from Tallahassee for the snapper that was abundant in the shallow waters. The crew began releasing a 600-foot troll line; it had barely had 100 feet in the water when a sea serpent struck the shark hook baited with mullet. The strike was so sudden and solid that it was initially thought the hook had snagged on a rock.

The "snagged on a rock" theory changed when the water erupted at the end of the troll line. The line was paid out for 300 yards and the serpent made a run to open sea. The ship, running at full steam, chased the line for four miles. Sensing it was tiring, they began to reel in the line. When it was in range, every gun on the ship was fired. One passenger fired a Winchester and scored a headshot. The serpent lunged at the ship then finally died. Crew and passengers combined to haul the carcass onto the deck by using a hawser and a capstan. The newspapers were careful to note it was the first sea serpent captured in the Gulf, although several had been sighted.

The serpent's corpse was brought back to the dock for public display. The snake measured 46 feet, 2 inches in length, and 72 inches around the body in the largest part. Although the body was snake-like, the head was described as being "spoon-bill-shaped" with a mouth resembling a shark's, but larger. The teeth were observed to be set at a 45° angle backward and it had a long forked tongue. The tail had multiple fins, 6 to 8 inches long. The color of the beast was brown, with a greenish back and a yellowish underside. The carcass was observed to be significantly slimy.

The Carrabelle Serpent is a textbook example of why some of these reports are difficult to believe. The article first appeared in the *Pittsburgh Daily Post* the next day—presumably, one of the passengers sent a message home as soon as the ship docked. It went national from there. By the time *The Wichita Daily Eagle* reprinted the news on September 18, 1896, less than a month later, the ship had gained another 70 passengers (passenger capacity was 50), and the monster had gained both three feet in length and the supernatural strength needed to have pulled the ship for 10 miles. A nice additional detail was added for dramatic effect, noting the strain on the *Crescent City* was so great that the timbers were sprung, and it was leaking when it reached the dock.

The description of the creature closely matches that of the frilled shark (*Chlamydoselachus anguineus*), one of two extant species of an ancient family of sharks, with fossil records dating back to at least the Late Cretaceous. Naturalist Ludwig Döderlein recovered several specimens in 1881 but never published his work, making ichthyologist Samuel Garman's 1884 description the first published report with the naming honors that go with it. His report, "An Extraordinary Shark," was published in *Bulletin of the Essex Institute* and placed the species in its own genus and family, naming it *Chlamydoselachus anguineus* from the Greek *chlamy* (frill) and *selachus* (shark), and the Latin *anguineus* for "eel-like." The frilled part of the name comes from six pairs of wide gill slits, so closely placed as to appear as a ruffle or frill. The shark's first gill slit runs continuously across the throat. Its jaws are terminal (at the end of the head, as opposed to underneath as in most modern sharks). The mouth contains over 300 trident-shaped teeth, arranged in about 25 rows set at an angle slanting backward. The pectoral fins are behind the gills, and the pelvic, dorsal, and anal fins are at the far end of the body, easily overlooked and adding to the eel-like appearance. Frilled sharks are rarely seen and are usually caught by deepwater trawlers off Japan and Australia, although examples have been captured or observed worldwide. The primary issue in such an identification is that Frilled Sharks rarely exceed six feet in length.

Necropsies of stomach contents indicate frilled sharks feed on cephalopods (octopuses and squids), other sharks, and bony fishes. One such bony fish is snapper, which may explain why the shark was near Dog Island, pursuing the same school of snapper as the fishermen aboard the *Crescent City*.

Garman suggests in the same article that the shark might be one of the sources of sea serpent lore and refers back to a similar report from 1880 where a 25-foot sea serpent corpse was caught in the nets in Muscongus Bay near New Harbor, Maine, by Captain Sylvetus W. Hanna of Bristol, Maine. Captain Hanna corresponded with Spencer F. Baird of the Smithsonian and reiterated that it had a flat head, shark-like gills, and an eel-like body. Heuvelmans scoffed at Garman's suggestion that a mere six-foot *C. anguineus* could be the source of sea serpent tales, suggesting that perhaps a larger form of another eel-like shark relative of the frilled shark also had survived. Teeth from extinct species in the fossil record suggest a larger member of the *Chlamydoselachae* family swam in millennia past, with a recent new species being named *Chlamydoselachus goliath* due to the notably large size of the teeth. In August 1907, a 12-foot carcass washed up in Sydney harbor. Australian marine biologist David G. Stead examined the remains and identified them as a frilled shark. If Stead was correct, the shark he examined was twice the length of any confirmed observed frilled shark sighting before or since. Perhaps Heuvelmans was correct in his suggestion that an even rarer, larger relative of the frilled shark had survived globally.

The year 1905 would add several additional frilled shark sightings along the Gulf Coast of Florida. In February, Captain Ernest D. Neeld of the fishing vessel *Pride* was returning from Old Tampa Bay to homeport in Tampa. As they were passing Gadsden Point, his mate noticed something coming toward them from Old Tampa Bay. Assuming the length (150 feet) and diameter (35 feet) is fisherman elaboration, the serpent reared its head out of the water with a loggerhead turtle in its jaws. Captain Neeld and mate Hugh H. Roberts agreed it had a head like a "Chinese dragon," an eel-like tail, large "side fins," and large teeth "shaped like corkscrews." A Chinese dragon in graphics of the time portrayed a roughly triangular head on a sinuous neck, which is closer to a frilled shark's anatomy than the Carrabelle "spoon-bill-shaped" head. The frilled shark's 25 rows of teeth could appear corkscrewed at a distance, particularly if a turtle was blocking a clear view. The beast circled the small fishing vessel and headed off, allowing the two men to downplay their fright by giving the story a folkloric finale by claiming the beast hit the coast so fast that it dug a canal across the point. There was also a report later that year. In late July, more than 50 locals spent most of a sunlit afternoon watching a serpent off the coast of Beacon Hill, a small town just above Cape San Blas. No additional specifics are provided, but the 35-foot serpent was spotted at a point only 40 miles due west of Carrabelle. Only a decade after Carrabelle's own encounter, it may suggest a breeding population along Florida's Gulf Coast.

Another report, three years later in 1908, suggests an even more extensive range for these theoretical larger frilled sharks. The July 15 issue of the *Miami Metropolis* included a front-page report that Turner Winfield from the town of Little River (annexed into Miami in 1925) encountered a sea serpent while on an outing in Biscayne Bay. Winfield, his wife, and a guest had taken a launch out for a short fishing trip. On the return home, the motor would not turn over and Winfield began to assume they would be forced to spend the night on the bay. He was able to work the boat in close to Belle Meade Island, where the Little River empties into Biscayne Bay. As he was making fast, he noted a disturbance in the water. Allowing for panicked observation, the physical description of the sea serpent could match that of a frilled shark: long fins just behind a head resembling an alligator, attached to a long slender neck, and a 30-foot length. The creature thrashed about in the water, which Winfield assumed was feeding. It headed out into deeper water, swimming like a snake before disappearing.

We don't know when this encounter took place—Winfield stated he had avoided reporting the incident out of fear of ridicule. He finally told the tale because others reported seeing the beast and the same water thrashing feeding behavior. Winfield's claim was not taken lightly—he was an early settler in Little River, which has the most extensive citrus groves in an area famous for its fruit and vegetable production. He was also active in the packing and shipping of produce, and one of a moneyed committee attempting to deepen Miami harbor.

However, the focus on the sighting and any additional details were hijacked by a local publicity hound named Charles Thompson. Captain Charley was a legend in fishing circles and a darling of the local media. If there was a record-breaking catch, a daring rescue, or any sort of adventure in the seas around Miami, Captain Charley was somehow involved. Captain Charley never caught the serpent, but he would gain national fame in 1912 for catching a 30,000-pound whale shark. The gentle giant would prove to be Captain Charley's meal ticket for a decade; he had the carcass stuffed and mounted for a cross-country tour.

The same year Captain Charley was taking on the whale shark, there was another sighting. This time it was on Cape Florida, the southern point of Key Biscayne, about 16 miles south of the mouth of the Little River. The 1912 sea serpent report didn't make the newspapers. But it was immediately brought back to mind by the Miami anglers in 1915 when the May 8 issue of the *Miami Herald* gleefully placed its tongue in its cheek and announced spring's official arrival—the first sea serpent sighting of the year had been reported.

On May 5, Payson Branch, a Miami musician, had gone on a fishing trip near Cape Florida, the southern portion of Key Biscayne. There, he and another musician saw a sea serpent in the shallow water covering the sand flats known locally for stone crabbing. As they neared the creature, it would hiss and move back toward deeper water. After several additional movements closer, it eventually slipped back into the sea.

They returned to the 12th Street dock but did not mention the encounter. Branch, the director of the Strand Theater's orchestra, eventually repeated the encounter to fellow musicians, one of whom leaked the story to the *Herald*. What caught the attention of the locals was an odd detail—the creature, over 50 feet in length, had a "fringe of pink whiskers" on the lower part of its head. Fisherman at the dock found this amusing, but also noted, that pink beard aside, the description matched previous serpent reports at Cape Florida in June of 1912.

It is important to note that Branch, a South Dakota native, had lived in coastal Florida for barely a year. He was such a novice that when a hunting party was assembled to look for his bewhiskered serpent ten days later, Branch nearly couldn't serve as a guide—he did not possess, and was unaware he needed, a fishing license. So, to a nautical novice, it may have been easy to mistake the gills of a frilled shark for "pink whiskers," keeping in mind that frilled shark's unique physiological features include the first gill slit continuing across the throat. The expedition came back empty-handed, fortunate for the serpent—in addition to a local motion picture cameraman, the fearless monster hunters carried enough ammunition to launch a military offensive. There were no further sightings of the whiskered serpent after the expedition, at least on the east coast of Florida.

In 1920, Cedar Key's economy was being rebuilt by a new sponge industry and the Greek sponge fishermen that dared to don 150 pounds of lead over a canvas and rubber suit and brass helmet to harvest sponge beds. So when a ship arrived in port with divers claiming to have encountered a sea monster, it was downplayed locally. There were other ports for the divers, most notably Tarpon Springs, and the economy did not need offended divers seeking a new port.

Word still got out, and as with the Carrabelle serpent, when the news reached other newspapers, the size grew with each newspaper's reprinting. The story was that while on the bottom, the divers encountered a "gigantic monster." The final exaggerated claims included a mouth as large as a door, eyes like two red lanterns, and a 100-foot-long body. The brief piece, carried in various papers, also noted that Spanish ships had been known to make at-sea trades of alcohol for sponges.

Stripping the hyperbole away, the divers, peering through the thick glass in the small window of the helmet, saw something unusually long, with an oversized mouth and something red extending out from the sides of the head. This may be the most tenuous identification of a frilled shark, but it is interesting to note that less than 300 miles of shoreline separate the four sightings on the Gulf Coast.

### Coelacanths in the Gulf of Mexico

The coelacanth is the most famous "living fossil," although it's technically not a living fossil—it's a "Lazarus species," a species presumed extinct until it was rediscovered. It was assumed the primitive fish had died off with the dinosaurs, 65 million years ago—right up until 1938 when a South African museum curator spotted one on a local fishing boat. The ancient fish symbolizes the cryptozoological quest to identify cryptids, essentially transferring their existence from folklore to zoology. It is no wonder the logo of the International Cryptozoology Museum in Maine uses the blue scaled fish to represent the museum and the quest.

So famous and captivating was the discovery of the dinosaur fish that unconfirmed sightings began to appear less than a decade after the first coelacanth's confirmed existence was announced. Soon, Forteans, the devotees of Charles Hoy Fort, who collected reports on all kinds of anomalies at the beginning of the 20th century, were claiming there were also coelacanths in the Gulf of Mexico. The cold, hard truth is that not one coelacanth has ever been trapped or seen in the Atlantic.

Henry G. Hoernlein was one such avid Fortean and a frequent contributor to the Fortean Society's magazine *Doubt*. *Doubt* included a section of summarized reports from contributors. Issue 27 in 1949 includes reports from California and Australia of fish with legs. That article dovetailed with Hoernlein's contribution. Whether it was Hoernlein's or *Doubt* editor Tiffany Thayer's doing is unknown, but the stripped-down version that ran simply stated that "A four-legged fish was caught near Tallahassee, Fla., January [1948]. Described as 4 feet long, with 3-inch legs. The Florida Wildlife Association could not identify it." Unfortunately, the story couldn't have been more inaccurate.

At the time, Florida Game and Fresh Water Fish Commission published *Florida Wildlife*, a magazine aimed at informing the public of changes to the law, promoting specific game lands, and announcing various topical news tidbits. One such small article appeared in a 1947 issue of *Florida Wildlife*.

> J. T. Wells, of Clearwater, gigged a mysterious critter that seems to recently be a candidate for Ripley's newspaper column. It had a catfish-like head minus the feelers, was four feet long and had four three-inch legs with four toes on each.

This description, based on the size of the legs with toes, almost certainly means Mr. Wells caught an amphiuma in the upper part of their size range. The amphiuma is a giant salamander, not a rare species but a rarely seen inhabitant of swamps and lowlands—the exact sort of environment one would go for gigging (i.e., hunting fish and frogs with a multi-pronged spear). It is but one of many reports of odd catches and amusing anecdotes that appeared in the magazine.

What made this one catch different was that the story caught the eye of someone in the United Press office in Tallahassee. It was sent out in abbreviated form across the wires for use as filler. The UP version edited out the location, inadvertently giving the impression a four-legged fish had been caught near Tallahassee (the only location in the piece being the UP dateline in the header). The article ends with "The Florida Wildlife Association says if you can name it, you can have it."

Although the wire service went to over 500 newspapers, it was not widely reprinted. However, one of the papers that did was the *Seattle Daily Times*. The article ran in the January 18, 1948, issue with changes—"gigged" was changed to "caught" and identified Wells only as a "Florida fisherman." Henry Hoernlein spotted the Seattle news brief. Neither Hoernlein in the Pacific Northwest, or *Doubt* editor Thayer in England, thought to review a map. The map would raise questions as to how a deep water marine fish was caught in Tallahassee, a land-locked city. This vague, heavily edited report was taken at face value by cryptozoologists as evidence that coelacanth range included the

Gulf of Mexico, a far cry from an oversized salamander caught in the swamps of Tampa Bay.

In December 1952, fourteen years after the discovery of the first living coelacanth, a second one was captured, generating another wave of press articles. *Science News Letter* of January 17, 1953, was one such article, with an interesting twist. Instead of focusing on the new fish, they noted the discovery "has reopened a mystery buried in files at the U. S. National Museum in Washington since 1949."

Ichthyologist Isaac Ginsburg of the U.S. Bureau of Fisheries (now the National Marine Fisheries Service) specialized in marine fish, particularly those in the Gulf of Mexico. One of his various job duties was answering correspondence from the public on marine life. (Co-worker Rachel Carson handled freshwater questions.) In 1949, Ginsburg received a letter from a woman in Tampa.

Enclosed with the letter was a fish scale that she wished to have identified. Ginsburg had no clue as to what species the scale came from. *Science News Letter* quotes Ginsburg as saying "This scale is like no other fish scale I have ever seen. It is not the scale of any of the several hundred known fish species of the Gulf of Mexico, and it is apparently of primitive structure." When pressed as to whether it could be a coelacanth by the magazine, Ginsburg was noncommittal. "It is not impossible that this is the scale of a coelacanth," Dr. Ginsburg said but added that the real significance was that the scale demonstrated, coelacanth or not, that there were still creatures in the Gulf waters to be discovered.

Ginsburg's near-obsessive level of care in cataloging and classifying species was even noted in his 1975 obituary in *Copeia*, the journal of the American Society of Ichthyologists and Herpetologists. He couldn't have seen a coelacanth scale in 1949 since the single known specimen was being displayed in South Africa. Therefore he couldn't commit to an identification.

Ginsburg never identified the woman who sent him the scale, but her letter included the circumstances of her acquisition of the scale. According to her, she bought scales from the fishermen to use in crafts. She purchased a gallon container of scales that she had never seen before. The scales were about 1½ inches in diameter, or roughly the size of tarpon scales but not similar in appearance to a tarpon. Ginsburg wrote for more information, but his letter was never answered. The scale is not among Ginsburg archives at the Smithsonian, but the mystery woman's reference to using the scales in craft projects offers an alternative reason Ginsburg might not recognize the scale.

In 1949, fish scales of all types were a secondary income source for the fishermen. Craft shops were buying scales in volume, bleaching them, painting them, and turning them into jewelry and artificial flowers. Scales from larger fish, such as tarpon (or coelacanths) curl when dried out, making them perfect for petals of roses and carnations. A May 1959 issue of *Mechanix Illustrated* profiled one store in the scale craft business, noting it had grown from humble origins in 1947 to an operation pulling in $50,000 a year (or $500,000 in current currency).

Tarpon fishing was a major tourism draw for decades in the Gulf States (they are now "catch & release" in most states). Since the tarpon is not good eating and doesn't take well to taxidermy, the souvenir offered was a photograph of the catch and a scale, dried and pressed, with your name, date, and the fish's weight and length added. Since the drying process takes several days, fishing captains would keep a stack of pre-processed scales on hand to use. Without the scale to examine, we will not know if Ginsburg didn't recognize the scale sent to him because it was an unfamiliar species, or because the scale had been flattened and sun-bleached, distorting the sample beyond identification with the equipment of the time.

This specious assumption of a Gulf colony of coelacanths was resurrected a few years later when University of Miami professor Donald P. de Sylva published an article in the May 1966 issue of *Sea Frontiers*. The article was about a silver pendant found in a village church near Bilbao, Spain. The pendant was a fish with a striking similarity to the coelacanth. It was discovered by chemist and eminent historian of technology Dr. Ladislao Reti, who believed it to be a century-old ex-voto, an offering in gratitude for divine intervention.

Reti's estimated age predated the rediscovery of the coelacanth in 1938. This prompted de Sylva to muse if it was evidence that coelacanths could be found where a Spanish sailor may have encountered one and later recreated it in silver. De Sylva specifically considered the possibility of somewhere between the Canary Islands, the Azores Islands, and the Strait of Gibraltar, or if a sailor had spotted a coelacanth near the Comoros Islands in an earlier expedition.

In 1965, a similar, but larger (14 inches) silver fish was photographed. This piece, with both paired and unpaired fins, proved the piece was unquestionably a coelacanth, identified by Dr. Jean Anthony of the National Museum of Natural History in Paris in his 1976 book, *Opération Coelacanthe*. German biologist Hans Fricke initially agreed in a 1989 article. Fricke, a coelacanth expert, consulted with José Manuel Cruz Valdovinos, an expert in South American silver at Universidad de Madrid, who thought the silver coelacanth was of Meso-American origin and probably dated back to the 17th or 18th century based on oxidation and the design of a hinge on the jaw. Additionally, Spanish artists of that time period added a stamp to mark their work while Native silversmiths were not allowed to mark their work. The inclusion of Meso-American silversmiths supported the theory that coelacanths were in the Gulf of Mexico since the Native population's craftsmen recognized them. Fricke already was suspicious. A Gulf of Mexico colony would be a significant find, but he and French marine biologist Raphaël Plante had scoured northern Spain and could not find any additional silver fish in churches or diocesan collections.

By 1997, Fricke was openly skeptical of both a Gulf coelacanth colony and the age of the silver fish. Cruz Valdovinos believed that errors in fin placement and head details suggested the silver piece was done from a photograph of the 1938 coelacanth. Fricke was even less generous, pointing out that the markings on the silver coelacanth show the representations of the white flecks of the coelacanth. Those marks were not on the original fish, nor were they discussed in print until the early 1950s. Fricke believed the silver coelacanth was created, at best, less than a decade before it was purchased.

In 2001, Fricke and Plante published an article in *Environmental Biology of Fishes* that debunked the silver coelacanth story once and for all. Professor Cruz Valdovinos had the opportunity to research the 1965 piece in greater depth. He noted the Reti pendant could not have been used as an ex-voto because fish were generally not used as ex-votos in Spanish churches.

He also noted the anatomical details of the piece were depicted realistically, lacking the stylized artistic embellishments of silver art pieces of the 16th and 17th centuries. Although the silver coelacanth would continue to appear as evidence of an undiscovered colony in the Gulf of Mexico, the silver had been permanently tarnished.

In 1994, J. Richard Greenwell, founding member and secretary of the now-defunct International Society of Cryptozoology, was a regular columnist for *BBC Wildlife* magazine. His March column recapped the rediscovery of the coelacanth and discussed whether the fish could be in the Gulf of Mexico, citing the silver coelacanth pieces as an example and offered a new story of unidentified scales in Tampa Bay.

Greenwell recalled a 1993 conversation with Sterling Lanier, a book editor turned full-time author (including a series of cryptozoology adventure novels). As Greenwell recalled the conversation, Lanier, before moving to Sarasota permanently in 1987, was a "snowbird," spending winters in Florida. He was at an art show near Sarasota circa 1973, selling the brass figurine of animals and fantasy creatures he sculpted as a hobby. At a nearby booth, another artist was selling jewelry made from fish scales, including a necklace made of what Lanier thought looked like coelacanth scales.

Lanier interviewed the artist, and in a statement given to Greenwell for the ISC's records, Lanier learned the artist searched the discard pile from a Gulf shrimp trawler for scales to use in his craft. One fish had caught his eye because of the "glitter" of the scales, and he removed them, one at a time from the carcass. Lanier was allowed to examine the scales and sketch them, but the necklace was not for sale. When Lanier moved to Florida permanently, his notes and sketches were lost in the move, as was his statement to Greenfield when the ISC disbanded in 1998.

Ironically, even with this dearth of details, when compared to the Ginsberg scale and the silver coelacanths, the Lanier account seems to be the most reliable piece of circumstantial evidence. That the scales were recovered from a shrimp trawler is telling. Shrimpers typically concentrate in an area primarily south of Tampa Bay, roughly down to Fort Myers, and that same area has a fair number of underwater caves and blue holes. This means the accepted habitat of the known coelacanth colonies, caves by day and active at night, seems to be replicated in the Gulf. And if Lanier's account seems viable, it could support the Ginsburg scale's identity as well.

The problem remains that there is simply no physical evidence. No photographs, no scales, no eyewitness divers, and no fish caught in nets. With the amount of diving and commercial fishing in the Gulf, the odds would seem to suggest that there should be more solid evidence than a handful of long-lost scales.

Still, Florida will forever have a connection to the coelacanth, even if none swim in the depths of the Gulf: the coelacanth was the inspiration for the *Creature from the Black Lagoon* (1954), released months after the second coelacanth was found. The film's underwater scenes, along with both sequels, were filmed in the clear waters of Floridian springs such as Wakulla Springs. *Revenge of the Creature* (1955) was shot entirely at Marineland and Jacksonville on the Atlantic coast. *The Creature Walks Among Us* (1956) opens with scenes from Fort Myers on the Gulf Coast, the same area the shrimp trawlers may or may not have snagged the coelacanthic source of the scales that Sterling Lanier observed.

**The Legendary Megalodon**

Mankind is terrified yet fascinated by sharks. If we're not killing them, we're studying them. And if we're not studying them, we're sensationalizing them in films and television. And no shark embodies this terrified obsession like the great white shark (*Carcharodon carcharias*). The largest predatory fish on Earth, great white sharks grow to an average of 15 feet in length, although even larger specimens, exceeding 20 feet and weighing up to 5,000 pounds, have been recorded. And then, there's the extinct ancestor that would give great white sharks nightmares—*Carcharocles megalodon*—the megalodon.

Megalodon is known by the public because of its huge fossil teeth, found nearly all around the globe in fossil marine deposits, a popular and profitable acquisition for divers and beachcombers. The teeth make it popular, but the scattered vertebra recovered help create a better picture for paleoichthyologists. Megalodon lived from about 66 million to 2.6 million years ago. With maximum lengths of 50 feet and weights up to 49 tons, it was the ultimate apex predator.

Of course, reports of alleged encounters with large, unidentified sharks have been proposed as evidence for the megalodon's survival—including Floridian waters. Author and mariner Thomas Helm accompanied a shark trawler out of St. Andrews into the Gulf of Mexico in the late 1950s. The first night out of the harbor, the ship anchored in 100 feet of water (17 fathoms) about 3-4 miles beyond Shell Island, the channel entrance to St. Andrews Bay. The mate, Dino, was a former sponge diver in Tarpon Springs. He recalled a time that he and another diver were on the bottom harvesting sponges on "Twelve Mile Reef," when a big shark came toward his friend. Standard practice was to hit the knock valve inside the helmet, which releases a blast of air—the bubbles scared off sharks and barracudas. This shark was so big as to be unimpressed by the bubbles. The shark circled a few times and then opened his mouth. Dino claimed the mouth was as big as two barrel hoops (roughly making the mouth 45-50 inches in diameter). The shark came in and swallowed Dino's partner. The shark started to swim away, the air hose and lifeline trailing behind him when he stopped, shook, and disgorged the diver. Whether the shark didn't like the taste of a copper helmet or the air from the exhaust valve was making him nauseous, the diver, soon to be nicknamed "Jonah," wasn't waiting to find out. Risking the bends, he shut off his escape valve and shot to the surface like a balloon.

Helm was justifiably skeptical of the story, noting that "fishermen and sailors have a built-in reputation for respecting the truth to such an extent that they are prone to use it sparingly..." And there's another reason to suspect Dino was playing the newcomer—"Twelve Mile Reef" is a fictional place, the setting of *Beneath the 12-Mile Reef* (1953), a story of competing families of sponge fishermen. It was filmed in Tarpon Springs and Key West, so it is possible Dino meant one of the areas where the underwater scenes were shot, but that is not how Helms repeated the tale, so it looks like Helms missed the reference entirely.

Helm's doubts did not remain long. An encounter the next morning gave him a reason to reconsider Dino's fish story. The ship was still anchored, due to some emergency repairs. Helm noticed what he first thought was the shadow of a cloud on the surface. It was not a cloud—it was a giant shark. Gradually it rose until it was only a fathom or two below the trawler. The giant fish lay motionless beneath the boat, and the crew tried to determine what species it was. The color on its back was black, dark brown, or dark blue, and proportionally, it most closely resembled a great white shark. The pectoral fins could be seen on either side of the trawler when the shark was directly under it, and when the tail was even with the stern, the head was just below the mid-ship boom. Using the ship's 60-foot length as a measure, the shark was 30 feet long.

The giant stayed under the boat for an hour, sometimes sinking deeper just to rise back to the surface. He was so close that occasionally the crew could feel his dorsal fin bumping the keel. The seamen initially identified the beast as the harmless filter-feeding whale shark, but the longer the shark sat there, the more doubts they had. The shark eventually swam off, leave a relieved crew to repair the engine and add a new story to their repertoires.

The whale shark (*Rhincodon typus*) is a slow-moving filter-feeder, the largest known fish species. The average whale shark has a length of 32 feet, varies in color, but bluish-gray or brown-black are common. By comparison, great white sharks average 20 feet, but there are unconfirmed accounts of great whites captured off Australia in the 30-foot range.

The Helm account joins a handful of stories of oversized sharks that are used as evidence of the megalodon. David Stead repeats a story from his early days in 1918 New South Wales of a giant ghost shark towing off heavily weighted crayfish pots. The Stead case is second-hand information and the alleged 115-foot size so outlandish that Stead has to justify it as exaggeration from the terror of the encounter.

Author Zane Grey, fishing in French Polynesia off Rangiroa in 1927, spotted an immense 40-foot shark, yellow with white spots, wider than his boat. Grey and his son saw a similarly colored shark in 1933 aboard a steamship leaving French Polynesia. In the case of the Greys, they declared it was not a whale shark while precisely describing a whale shark. Loren Grey further noted a yellow tint in the water before the second sighting, which marine biologist Ben S. Roesch suggests was a "cloud" of plankton, the exact sort of thing to attract a filter-feeding whale shark.

Zoologist Karl Shuker, who does believe there may be living fossils still to be rediscovered, is not a strong advocate of the megalodon's survival. He notes in his 2016 book that there are no reports of megalodon-sized sharks in historical records where one might expect them, such as the shark frenzies near whaling ships as they butcher their catch. Nor have there been sightings of cetaceans with megalodon-sized scars.

Roesch goes a step further, pointing out if the megalodon were still alive today, it would inhabit the same shallow waters as the great white shark. Considering the proliferation of people along the coasts, it is highly unlikely that the megalodon could be missed by surfers, swimmers, and boaters.

Shuker considered the megalodon one of the more terrifying of cryptids. He doesn't specify if the terror is derived from the thought of a 45-foot killing machine with serrated teeth up to 7 inches long, or from the sheer amount of contention in cryptozoological circles over whether the megalodon has actually survived.

**The Miami Beach Sea Scorpion**
March 11, 1959, was a typical day at work for Bob Wall. Wall was an underwater tour guide aboard the glass-bottomed boat *Comrade II*. His job was to dive beneath the boat and point out coral and fish for the passengers, while Captain William Wood handled the vessel and narrated. Two miles off Miami Beach, in 35 feet of water, it was business as usual. Wall noticed a small cave in the coral and looked in to see if there was something he could flush out for the tourists above. Instead, something looked back at him.

Wall had been diving in these waters for over a decade and had never seen anything like it. It was five feet six inches long with a long cylindrical body, no tail, with eight legs that lifted it three feet off the ocean floor. The legs were "hairy like a sea spider." The head was pointed and included brown spotted eyes the size of silver dollars on stalks. Initially, Wall thought he was looking at an oversized spiny lobster, but as he noticed differences to the lobster, the creature started moving toward him. Wall immediately returned to the boat. Captain Wood, hearing his story and noting how shaken the diver was, dropped a marker buoy.

Returning to the dock, Wood contacted the Miami Seaquarium to borrow an underwater camera. The next day, however, the seas were so rough that *Comrade II* couldn't anchor near the buoy. As news spread Miami, people were perplexed. The *Herald's* outdoor writer and resident marine expert, sight unseen, immediately suspected an octopus. Wall insisted it had no tail so it wasn't a lobster, and it had a pointed face, so it wasn't a crab.

Captain Gray of the Miami Seaquarium was interested enough to send out his own divers to assist. He believed Wall's initial impression was correct and it was an unimaginably large spiny lobster. Gray noted an 18-pound spiny lobster had been recently captured off Belize, which at four feet, was twice the average length—impressive, but still smaller than Wall's sighting. But huge lobster or not, it could be a new attraction at the Seaquarium.

By March 15, even the Seaquarium had given up. Rough water had stirred up the bottom, Assuming someone was able to dive down and investigate, the creature was gone and so was the interest of the local newspapers. The *Miami Herald* barely covered the story, but the *Miami News* reported the story as ongoing news, if occasionally with its tongue firmly planted in its cheek. Bob Wall was one of Miami's earliest proponents of scuba diving, and the Seaquarium's involvement gave the story some legitimacy, even if some local wag named the creature "Monty the Monster," which the *Miami News* adopted as well. Other newspapers that picked up the AP story treated the report with less seriousness. *Fate* magazine's July 1959 also carried the story, based on the *Miami Herald* coverage.

The story quickly faded into obscurity until Fortean researcher Mark Hall released one of his popular self-published books in 1991. In *Natural Mysteries: Monster Lizards, English Dragons, and Other Puzzling Animals*, Hall was examining reports of freshwater octopuses in Oklahoma before somewhat arbitrarily deciding that these accounts were living fossils known as eurypterids. Dubbed "sea scorpions" because of a spine-like appendage on the tail, the eurypterids are more closely related to modern arachnids than scorpions. They flourished in warm shallow water from 460 to 248 million years ago.

Hall used the eurypterid as a catch-all for any multi-legged aquatic cryptid he couldn't immediately recognize. Other potential eurypterids or a cryptid evolved from them, at least according to Hall, included the giant tentacled Bahamian lusca, and a creature pulled from the cold depths of the Gulf of St. Lawrence.

Shuker (1995) doesn't see how such a wide variety of differing physiologies could all be from eurypterids but doesn't dismiss the notion that some of the sea scorpion's descendants might survive. As for Bob Wall's encounter off Miami Beach, Shuker tends to think the Miami Seaquarium was right—a massive spiny lobster and a startled diver combining for a new tale of the sea.

Hall's most lasting contribution to the Miami Beach cryptid may be the name. Although the *Miami News* refers to the mysterious creature as "Monty," Hall claimed the cryptid was nicknamed "Specs" for its eyes. His only source for the Miami story is a UPI newswire story from the *Miami News*, which does not mention either moniker. This suggests Hall is the origin of the Specs name.

In 2014, a fisherman reeled in an 18-inch long "shrimp" caught in the Indian River near Fort Pierce. The Florida Fish and Wildlife Conservation Commission posted several photos on social media, and the crustacean was a brief viral internet sensation. The FWC, attempting to identify the mantis shrimp from photos, could only narrow it down to a type of "mantis shrimp," a kind of stomatopod, only distantly related to shrimp.

These solitary creatures spend most of their time hiding in rock formations and have the correct shape, number of legs, and eyes on stalks, as described by Wall in 1959. The issue remains size. One of the largest stomatopods is the smooth mantis shrimp (*Lysiosquilla scabricauda*), which is found in Floridian waters and can get as large as 9-12 inches, big enough to be the type caught in Fort Pierce but still a far cry from Bob Wall's 5-foot crustacean. So there may be a larger species of crustacean off the coast of Florida, or Bob Wall's panicked calculation was a little exaggerated. The answer may be a fishhook away.

## Goblin Sharks

The living fossil known as the goblin shark (*Mitsukurina owstoni*) was first discovered off the coast of Japan in 1898. It is one of the most bizarre-looking sharks known, pinkish-grey in color, averaging 10-12 feet long and with a distinctive flattened and elongated snout above an extendable jaw of jagged, nail-like teeth. The name "goblin shark" is a literal translation of the Japanese name *tengu-zame*, a tengu being a Japanese spirit depicted with a long nose and red skin.

Similarities to fossil goblin sharks from the Upper Cretaceous (100 million years ago) prompted a century-long debate over whether the goblin shark should be considered the same genus *(Scapanorhynchus)* as its fossil ancestor. The discussion was academic for much of the world since, like the coelacanth, the few examples captured were limited to a small geographic area, Japanese waters. As commercial fishing went into deeper waters, more specimens were discovered in more diverse locations. The capture of a goblin shark in the Gulf of Mexico in 2000 suggested the species had a global range.

The first goblin shark caught in the Gulf was the night of July 25, 2000, by commercial red crab fishermen 550 fathoms deep, 34 nautical miles southeast of Redfish Bay, Louisiana, or 108 nautical miles southwest of the Florida-Alabama border. The shark was female, with an empty stomach and no evidence of carrying pups. Photos were taken, the jaws were saved as a trophy, and the rest discarded. Estimates from the photographs suggest the shark was nearly 18 feet, the largest ever caught.

A second goblin shark fared better. This shark was caught April 19, 2014, off Key West by Carl Moore, trawling for royal red shrimp 15 miles south of Key West at 285 fathoms. Moore released any bycatches, including the 18-foot-long goblin shark thrashing around on the deck of his 105-foot boat. Moore and his crew were able to get a rope around the end without the teeth and hoist the shark over the rail with a winch.

Both sharks were in the 18-foot range, well above the current average for the admittedly understudied species. Perhaps larger examples are found deeper than the fishing industry has delved, or simply regional environments are conducive in the Gulf. Either way, the goblins are getting bigger.

# CHAPTER 2
## Sea Monster Fatalities

Other than accounts in folklore such as mermaids luring sailors to their doom, or the hapless fisherman that Nessie killed before St. Columba stopped his reign of terror, there are very few marine cryptid encounters that involve fatalities. Florida has the dubious distinction of having two alleged sea serpent sightings that led to tragedy.

### The McClatchie Tragedy

On 4th St North in St. Petersburg is the Cathedral Church of St. Peter. Included within the venerable Gothic Revival architecture is a particular stained glass window of "Christ Coming to Mary and Martha" that was placed in the nave in 1923. It may be the only stained glass window dedicated to the victim of a sea monster.

Dot McClatchie and Mary Buhner had graduated from St. Petersburg High School together on June 8, 1922. The two best friends had dominated the high school athletics, Dot lettering in basketball, track, and swimming, and Mary lettering in swimming and serving as captain of the swim team. The two planned to attend the same college in the fall.

On June 17, they spent the day at the beach, attending an afternoon wedding. Afterward, the two athletes decided to swim out to the channel buoy a mile out in Tampa Bay, then dress for dinner. Reaching the buoy and preparing to head back, Dot suddenly jerked in the water and screamed that something had bitten her. She jerked again and screamed her foot was gone. Mary swam over and could see both legs were attached, but blood spread across the water from a significant leg injury. Dot went into shock and stopped swimming. Mary attempted to tear a piece off her suit to make a tourniquet but was unsuccessful. Mary tried to do a sidestroke while holding her friend's head above water but was making limited progress back to shore. The waves were so choppy that no one could see the girls were in trouble from the shore.

Mary began screaming for help, and a vendor on the pier finally heard her after 20 minutes. He ran to the yacht basin and found someone with a boat.

They raced out to assist, but the two were only visible between wave crests, slowing down the rescue attempt. When the rescuers pulled Dot into the boat, she was unresponsive. The wounds had stopped bleeding—she had lost so much blood that her heart could not pump enough to send it to her legs (hypovolemic shock). Mary had trained as a lifeguard and performed CPR as the boat rushed back to shore. An automobile was commandeered to drive her to the hospital, where she was pronounced dead.

A postmortem examination revealed the extent of the damage. Something had inflicted two deep wounds, one on her thigh and one on her calf. The wound on Dot's thigh was 14-inches long with 13 separate tooth marks. The other wound extended three-quarters of the way around her the calf, just below the knee. Her femoral artery had been severed and the injury extended partly through the bone. Nothing Mary could have done would have saved her friend, but Mary would never accept that fact.

Dorothy McClatchie was buried in Saint Bartholomew Episcopal Cemetery in St. Petersburg. Her gravestone is a broken column, symbolizing a life cut short. The base of the stone lists her name and birth date. Her death is noted as "killed while swimming in Tampa Bay - June 17, 1922." Dot was 18 years old.

The funeral received minimal coverage compared to the debate over what animal was responsible for the attack. A shark was the obvious suspect, but a sea monster was, of course, suggested, as were a porpoise, a jewfish (*Epinephelus itajara*), a swordfish, and a sawfish, before a barracuda became the accepted culprit. The selection may have been more damage control than forensics. Summer tourism was St. Petersburg's lifeblood, and a shark, which could come close to shore, could cause a panic. A barracuda could be more easily dismissed as a deep-sea beast, making Dot's death an unfortunate accident that wouldn't impact visitors to the beaches like a shark attack would.

Australian surgeon Victor Coppleson revisited the matter in his book on shark attacks. His conclusion was the damage done to Dot McClatchie was consistent with an attack by a six-foot shark. That size suggests a Bull Shark (*Carcharhinus leucas*), which is one of the most frequent culprits in attacks on humans in Florida waters.

Whether Mary could have put Dot's death behind and move on with her life, the simple fact was she was never given the opportunity. She was a celebrity, touted as a heroine as the newspapers debated which terrible beast had killed her friend. The city gave her a commendation. The Kiwanis started a college scholarship fund. Eleven months later, not only was she was still identified as the stalwart school chum who bravely tried to save Dot McClatchie, she was awarded a Carnegie medal for heroism, meaning the story went national again. In 1925, her portrait was given to her high school in honor of her heroism.

Among all this accolade, there is one obvious thing missing—Mary herself. She talked to the *St. Petersburg Times* the day after the attack while she still in shock and possibly medicated to help with the muscles she tore keeping Dot afloat. After that, she avoided the press and the accolades. Constantly reminded of her friend's death, she simply refused to cooperate. Instead of going to college as she and Dot had planned, she married Fred York, a dentist she had been casually dating. Even the name change didn't help. The newspaper just referred to her as Mary Buhner York. The scholarship money that had poured in totaled nearly $2000 ($30,000 in today's money) was used to pay for the window in Dot's memory.

Fifteen years after the tragedy, Mary broke her leg in a fall while hiking. It was covered in the newspapers but the fact she had to be carried down the mountain and placed in a truck to drive to a hospital was secondary to the fact that she was a Carnegie Medal winner for the sea monster encounter.

Mary and Fred York divorced in 1943. Mary immediately moved to Knoxville, Tennessee, where an old family friend had relocated to run the local tourist bureau. She became the tourist bureau's secretary, living out the rest of her life in anonymity, far from the nightmare sea serpent of her youth.

### The Pensacola Pass Plesiosaur

One of the most notorious sea serpent reports in cryptozoology annals came to the attention of researchers in 1965. While there is obviously no shortage of sea serpent sightings in Florida, the area around Pensacola had been quiet. Other than Thomas Helm's 1943 possible cryptid pinniped sighting 100 miles away in St. Andrew Bay in 1943 (Chapter 12), there is no record of sightings along that section of coast. So when the May 1965 issue of *Fate* magazine hit the newsstands, cryptozoology researchers were taken by surprise. Author Edward Brian McCleary claimed he was the sole survivor of a sea monster attack in 1962.

In the version McCleary recounted in *Fate*, he was 16 years old when another teen named Eric Ruyle invited him to go skin diving later that Saturday morning, March 24, 1962. McCleary agreed. Rounding out the group was 17-year-old Warren Salley, 14-year-old Larry Bill, and 14-year-old Brad Rice. McCleary admitted he was completely unfamiliar with the dive site—

the wreck of the *Massachusetts*. The *Massachusetts* was a WWI battleship, built in 1893 and scuttled by the Navy in 1927 to be used as target practice. Lying in 25-30 feet of water, part of the ship is still exposed to this day. No longer of use to the Navy, it had become an artificial reef and the destination of the increasingly popular new hobby of skin diving.

Their destination was Fort Pickens State Park (now maintained by the National Park Service) at the western tip of Santa Rosa Island. An hour later, they were parked near the fishing pier and preparing to launch a seven-foot Air Force life raft, equipped with a drift anchor and oars. McCleary noted they had climbed to the top of Fort Pickens and could see the wreck, two miles off the coast. He observed the water was cold as they launched the raft.

They took turns paddling the raft. Salley was paddling when Bill noticed the wreck was now on the left. But when they started out, it was on the right. In other words, the current was pushing them to the west. Salley began complaining about tiring out because of the rough water. McCleary noticed that whitecaps had formed, and the sky was clouding over as the wind picked up.

They decided to turn back, but the wind and waves were pushing them into the Bay channel. Ruyle, McCleary, and Salley jumped into the icy water and began kicking behind the raft while Bill and Rice took the oars. But the tide was just too strong. They climbed back into the raft, shivering and cramping from the cold. The waves were so high they had to hold the sides. The sky continued to darken, and they could see small crafts heading to port in the distance. They tried to wave down a passing Chris-Craft with no success. Bill began getting hysterical.

McCleary pointed out a buoy about a mile in the distance, and they began paddling toward it. The plan was to hook it with the drag anchor. As they neared it, the waves were threatening to swamp the raft—only the inflated sides kept them afloat. They neared the buoy. But before he could toss the drag anchor, the waves lifted the buoy up from its mooring and a riptide formed beneath. They looked at the 20-foot-tall, red-metal behemoth as McCleary stood up and hurled the anchor like a lasso at the buoy. Before the anchor reached the buoy, the raft was caught in the current and dragged straight toward the bottom of the buoy. McCleary yelled for everyone to jump as the buoy came down off the wave, smashing into the raft.

Salley spotted the raft resurfacing. McCleary and Ruyle reached it first and were able to flip it over. They climbed back in as a cold driving rain began. They watched helplessly as the current pushed them past the *Massachusetts*. McCleary noted the wheelhouse sticking out of the water and wind roaring through the windows, making a sound like a siren.

The five teens lost track of time, but eventually, the rain tapered off to a fine mist and the sea subsided to a dead calm. A thick fog rolled in and McCleary noted it was unnaturally silent; the silence just reemphasized their predicament. Rice began to panic. They tried to calm him down and decided all they could do is sit and wait. The fog limited them to 25 feet of visibility. The water, which had been notably cold at the beginning of the trip, was now unusually warm, even for summer, yet this was only March.

Bill suddenly bolted upright and said he heard a boat. No one else heard anything, but the air suddenly became thick with a sickening smell of dead fish. Forty feet away, a tremendous splash generated waves that reached the raft and broke over the edge. They heard another splash, and through the fog, they could see a shadowy form that looked like a telephone pole, 10 feet tall with a bulbous shape on the top. It stood there for a moment, then bent in the middle and dove underwater.

A high-pitched whine broke the silence and the boys panicked. The five of them put on their swim fins and jumped into the water, which was covered in patches of crusty brown slime. McCleary noted a slight current that he hoped would lead to the shore. Instead, they decided to try to reach the *Massachusetts*. Ruyle and McCleary took the lead, with the other three close behind. Whatever they had seen, they could now hear, hissing and splashing behind them. The fog was clearing but the water was getting rougher. It began raining again and the water was getting colder again.

McCleary was beginning to cramp, so he started swimming with slow, deliberate strokes, more concerned with staying afloat. Ruyle was still nearby, and they would call back to Rice, Salley, and Bill. As they swam toward the wreck, they suddenly heard a scream and Salley yelled that the monster had gotten Brad, but his yells were suddenly cut off by a short cry. Bill caught up with Ruyle and McCleary. The only sounds were the ocean and thunder. McCleary slipped in a fugue—unaware of the ocean depth or what was out there, and imagined sinking peacefully to the bottom. The pain in his legs snapped him back to reality and he realized Bill had vanished. He and Ruyle dove under looking for him, but there was no sign of him. Ruyle grimaced and also went underwater. He was cramping up badly. McCleary had him wrap his arms around him and began swimming. They struggled through the water for several hours. Night had fallen and the two struggled onward in the dark, waves breaking over their heads.

Just about when McCleary was about to give up, a lightning strike lit up the sky and they saw the silhouette of the *Massachusetts*. A wave broke over them and the two separated. Another bolt of lightning showed Ruyle ahead of him, swimming toward the ship. The creature surfaced next to Ruyle. McCleary noted the long neck and small eyes. The mouth opened, and it dove on top of Ruyle. McCleary screamed and swam past the ship. He didn't remember swimming the two miles from the wreck to the shore. He slept in an abandoned watchtower until the morning when fishermen spotted him.

Bernard Heuvelmans briefly mentioned McCleary's story in the 1968 English version of *In the Wake of Sea Serpents*, unwilling to commit to the story's authenticity because he had only been made aware of it just before releasing the English edition. As such, he was reluctant to speculate. Other authors were less cautious, using the *Fate* version as if it was the only source available.

The *Fate* article embellishes details when compared to the newspapers of the time that anxiously covered the search for the boys and McCleary's rescue. Allowing for exaggeration and hyperbole to make the story more marketable for *Fate*, there are factual issues as well. Charts of the harbor for the time confirm a red buoy nearby; just 200 yards off the wreck, designated "WR2." But also note the buoy had a bell and flashing red light, making the location of the wreck reasonably easy to detect in fog, rough seas, or darkness. All other red buoys would be on the starboard side of the channel. The nearest buoys would be the port side markers, which would be black in color. The nearest starboard (red) buoy without a light or bell would be 1.5 miles away on the far side of the channel, and the port buoy on the opposite side had a light. In fact, it would be difficult for the raft to become lost unless they were pushed across the channel into the shoals on either side of Pensacola Pass, which would require a direction perpendicular to the current and would make land still visible during the daylight hours.

So what actually happened to Brian McCleary and his friends that day?

A simple review of the 1962 newspapers confirms that McCleary was involved in a tragic trip into Pensacola Pass where four of his friends disappeared, but there are notable differences between the contemporary newspaper accounts and the version in *Fate*.

Brian McCleary's mother was in Fort Walton Beach Hospital for a series of tests when her son's friend, 16-year-old Eric Ruyle, came to the hospital to ask if Brian could go spearfishing the next day off Fort Pickens with a group of friends. What Ruyle didn't stress was how far off Fort Pickens that Ruyle and his friends were planning to go.

Saturday, March 24, 1962, was a perfect day for snorkeling. Clear skies, temperatures in the mid-60s°F, and no rain in the forecast. More enticing to Brian McCleary was their destination; he had only lived in Florida for a year, and while already an avid skin diver, he had never visited a shipwreck. His mother would later tell the newspaper that he had never been in a boat, let alone a raft. So it appeared that none of the teens had a sense of the potential risks.

Even today, the *Massachusetts,* although in shallow water, is not considered a dive for amateurs or novices. The ship was scuttled in a location selected for military target practice, its future popularity among divers was not considered. It is in the Pensacola Pass, an outlet that connects Pensacola Bay with the Gulf of Mexico. The location is prone to strong currents and rough surges. The adventure was doomed almost from the start.

The version of events in local newspapers was based on interviews with McCleary, and the difference between the two versions is striking. In the original newspaper version, the teens had barely paddled a quarter-mile from shore when swells started threatening to swamp the raft. They decided to head back to shore but discovered they were caught in a strong tide that was pushing them out into the Gulf of Mexico. The raft passed a channel marker buoy that the teens attempted to hook onto with the drag anchor. It was unsuccessful, but it also confirmed they were being pushed out into the open sea. The boys debated swimming to the buoy, but within 15 minutes, it was already a half-mile away. Panicking, the boys decided to abandon the raft and swim to a buoy.

The plan was to stick together and help each other reach the buoy. The plan fell apart as soon as they hit the water. The current was too strong and the boys were exhausted from attempted to paddle back to safety. Salley and Rice immediately became separated in the swells. Ruyle soon developed muscle cramps. McCleary and Bill attempted to hold Ruyle up, only to have Mc-Cleary develop leg cramps as well. McCleary tried to get Ruyle to hold on to him but told him he didn't think they were going to make it. Ruyle told him they were going at least going to try. Another swell separated McCleary and Ruyle. In this dinosaur-free version of events, McCleary told searchers he didn't actually see any of the four go down.

McCleary was a strong swimmer, probably the strongest in the group. He kept his head and realized that to battle against the current was futile. Using the buoys as landmarks, he began swimming at an angle, cutting across the current rather than battling it head-on. Using only his arms until his legs began to loosen up, he swam for hours. Feeling the tide changing direction, he allowed it to push him toward land. He reached the shore after dark. He had been in the water for more than five hours. He staggered onto dry land but couldn't find any buildings to summon help. Exhausted, he fell asleep against what he thought was the Coast Guard watchtower on Santa Rosa, seven miles east of their starting point at Fort Pickens.

McCleary was found at 6:45 the next morning by spearfishermen. He learned that the Coast Guard had spent most of the night searching for them. He had been sleeping against an old gun emplacement at the abandoned Fort McRee on the eastern tip of Perdido Key, almost directly across the Pass from where they had started. The fishermen took him to a nearby trawler that radioed for a Naval Air Station helicopter. The copter that took him to the hospital for observation was piloted by Major Ralph Ruyle, Eric Ruyle's father.

The raft had washed up further west on what is now called Perdido Key Beach, roughly where the Beach Colony Resort now stands. It had not capsized, swamped, or been crushed by a buoy. It still contained five spear guns, five face masks, three fins, a knife, five pairs of shoes, five towels, three hats, and a cap.

By Monday, as reports from the search ship began to filter in from the weekend, the press was referring to McCleary as the sole survivor. With McCleary's description, the search area was narrowed. Pensacola Naval Area Station joined Escambia County Search and Rescue as 20 boats dragged the area around the Pass. The search was called off Friday afternoon. Navy aircraft crews were ordered to keep a lookout for bodies, but there was little doubt it would be a recovery, not a rescue.

Frank Neilsen, one of the shore-based searchers, had not given up. The ocean had been particularly rough the night of March 30th into the morning of the 31st, and Neilsen hoped perhaps it would be enough to push one or more the remains back to shore. He was correct. Shortly after noon, the body of 14-years-old Brad Rice washed ashore near Fort McRae. He was still wearing swim fins.

A memorial service for Eric Ruyle, Warren Salley, and Larry Bill was held that night in a chapel at Eglin Air Force Base. Their remains were never recovered. Brad Rice's funeral took place on April 2, followed by interment at Fort Walton Beach's Beal Memorial Cemetery.

This version, pieced together from contemporary newspaper reports, paints a significantly different version than the tale in *Fate*. Where McCleary's 1965 version involves a squall and fog, no 1962 newspaper report mentions such weather conditions, including McCleary's firsthand account. Additionally, the daily weather report for March 24, 1962, from the Sherman Naval Air Station clearly showed the 1965 account in *Fate* is incorrect. At the time the boys launched the raft, the wind speed was 13 miles per hour and diminishing. The visibility was 5 miles. There was no precipitation recorded for the day. The Naval Station reported fog, but based on hourly temperature and dew point records, fog could only develop briefly in the early morning; the sustained winds of 8-to-12 mph winds would be problematic for any fog to last very long, let alone until the afternoon when the winds increased.

McCleary knew the newspaper reports of the day and the *Fate* article had significant differences, not the least of was a glaring lack of sea serpents in the original account. His preemptive answer was that the director of the Escambia County Search had initially advised him to keep quiet about the sea monster and let the matter fade away as an unfortunate tragedy at sea.

The *Fate* article itself was an anomaly. McCleary attempted no publicity before or after this article and avoided interview requests. Among the readers of McCleary's story in *Fate* was cryptozoologist Tim Dinsdale, famous for his 1960 film footage of something in Loch Ness and his 1961 book on the beast, *Loch Ness Monster*. Dinsdale was working on his second book, *Leviathan*, in which he was including reports of other lake and sea monster sightings, a way to bolster's the Loch Ness creature's legitimacy by showing it was not an isolated case. Dinsdale was struck by how similar the description of the Pensacola creature was to another case, one he had mentioned in the 1961 book—a 1910 sighting in the Bay of Meil off the Orkney Islands by W. J. Hutchinson. He wrote to the magazine, and the editors forwarded the letter to McCleary. McCleary wrote back, including additional details, some of which conflicted with the particulars in *Fate*, let alone the actual event. The neck, brownish-green and smooth, was now 12-feet long, adding two feet to the description in the article. The head, formerly described merely as a bulb, now was like that of a sea turtle, only more elongated, with teeth. The eyes were green with oval pupils, and there may have been a dorsal fin. And now McCleary claimed he stayed most of the night aboard the wreck of the *Massachusetts* and swam to shore in the early morning.

McCleary included a drawing of the creature that showed a long-necked, dinosaurian creature surfacing in front of a channel buoy. The curvature of the neck is reminiscent of the widely reproduced "Surgeon's Photograph" of the Loch Ness monster. More telling is the buoy, showing a lateral marker buoy with a "9" written on it. Nine, being an odd number, would be on the port side of the channel heading to shore, black, not the red as described by Mc-

Cleary. A review of the Pensacola Pass marine charts of the time, the number 9 buoy was not lit, meaning it was a "can buoy," flat-topped and cylindrical in shape.

So the question remains why are there multiple versions of the story?

The answer may simply be the 2¢ a word that *Fate* paid writers. McCleary had gone on with his life. He graduated high school in June of 1963 and attended a local Junior College before transferring to Louisiana State University for the Fall semester of 1964. He met a local girl, a sophomore named Paula. By August 1965, they were engaged, and they married on Labor Day weekend. By the time he was engaged, his story had run in *Fate*. The $40 (more than $320 in 2019 dollars) would have proven helpful for newlyweds setting up an off-campus home.

McCleary found work and attended LSU part-time while his wife continued school full-time (underwritten in part by legacy scholarships provided to alumni children). She graduated in May 1966. McCleary graduated in 1968, and the couple moved to Connecticut so he could start a career, somewhat ironically, as a life insurance claims adjuster. They moved with some frequency. He continues to maintain a low profile, unwilling to discuss the matter, which has left him and the incident undefended against claims of fraud, exaggeration, and other salacious nonsense.

The *Fate* story is essentially a fictionalized version, but four families lost sons that day in 1962. They are painfully aware of how real the tragedy was. A 14-year-old boy is buried in Fort Walton Beach, and three sets of parents did not even get that much closure. There is no need for a sea monster to make it worse.

# CHAPTER 3
## Sea Monsters of the Gulf

The Gulf of Mexico basin connects to the Atlantic Ocean through the Florida Straits between the U.S. and Cuba, and with the Caribbean via the Yucatán Channel between Mexico and Cuba. It's big and isolated—the perfect place to encounter something big and unknown. In 2019, that statement came true. Dr. Edith Widder, the lead scientist of the Ocean Research and Conservation Association out of Fort Pierce, captured video of a 10-to-12 foot juvenile Giant Squid (*Architeuthis dux*), which can grow to 40 feet. The footage was shot in the Gulf of Mexico, roughly 250 miles west of St. Petersburg, the first giant squid ever recorded in U.S. waters. To do so, Widder and her team imitated the bioluminescence of a distressed jellyfish, the same technology they had used when they filmed the first-ever giant squid off Japan's Ogasawara archipelago in 2012.

So, having established monsters do dwell in the Gulf, what other cryptids might still be waiting to be discovered?

### The Cuban Leviathan
Nicolas-Joseph Thiéry de Menonville was not a happy man on the morning of August 27, 1777. His ship was sailing along the northern coast of Cuba; its destination was Saint-Domingue, the French colony on Hispaniola that would later be known as Haiti. Thiéry de Menonville was a French botanist who had just left Veracruz, Mexico, in disguise. His mission of industrial espionage for the French Government was more dangerous than it sounds today—he landed in Veracruz, then sneaked into the neighboring state of Oaxaca in disguise. His goal was to steal a breeding colony of cochineal insects, a small beetle. Once dried and ground up, the powder produces a brilliant scarlet-colored dye. Ounce for ounce, this dye was Spain's most profitable export from the New World, and Spain was aggressively protective of its monopoly, hiding the source for centuries. But after two centuries of keeping the source secret, it had gradually emerged that the tiny insect was the source.

Thiéry de Menonville had discovered that acquiring the insect was only part of the problem—the insects were parasites that lived on only one plant, the nopal, the Mexican Spanish name of the *Opuntia cacti*, known in English as a prickly pear cactus. With samples of both carefully camouflaged within a collection of uninteresting plants, he planned to establish a plantation of the cactus in a *Jardin du Roi* at Port-au-Prince. Assuming of course, that he could keep the insects and plants alive on the voyage from Vera Cruz, across the Yucatán Channel, around Cuba and back to Port-au-Prince. But first, he had to get out of Mexico. That was proving more difficult than expected. By late May, he was frantic to leave before his cacti began growing too big for their packing crates. Then he learned of a ship going to Cap-François, the commercial heart of the French colony, but on the opposite side of Saint-Domingue. The destination was not his first choice, but at least he'd be out of Spanish territory.

His journey notes were published posthumously in 1787 as *Voyage à Guaxaca*. The adventure starts as something out of an Ian Fleming story—hobnobbing with pompous Spanish authorities, false identities, disguises, contraband plants in hidden cases, and narrow escapes. The captain decided to sell his cargo in Havana instead, then waited for a new shipment heading to Port-au-Prince. When the vessel finally departed, Thiéry de Menonville was starting to lose his plants. By August, the narrative had a more self-piteous whining nature—an incompetent navigator, days of severe storms alternating with days becalmed, encounters with British ships looking for Colonial rebels, and a captain still in no hurry to reach Saint-Domingue. And all of this chronicled in his journal against his anguished updates on his plants. If the plants died, so did the parasitic insects, making the arduous journey for naught. So, when the morning of August 27, the botanist had a sense of relief as the ship finally made progress—right up to when a lookout spotted something in the water ahead of the ship. The captain's first thought was that it was a reef, 180 feet long and mostly white. The captain thought it could be a capsized ship. Thiéry de Menonville, based on the sharks and birds circling it, was convinced it was just the carcass of some deceased sea creature.

To Thiéry de Menonville's horror, the captain decided to investigate, creating yet another delay. The captain gave the order to come within 100 feet of the object. Once they were within 300 feet, the stench of decomposition was so overpowering that there was no question it was the decomposing corpse of something huge that had been dead for some time.

It was indeed a massive carcass, surrounded by a "dazzling white" greasy foam, creating the mass along the surface initially mistaken for a reef. Thiéry de Menonville noted the main mass was 90 feet in length with seven or eight adjacent small pieces of 12-to-18 feet. The breadth was over 40 feet wide, and it sat with three feet exposed above the water, but with another 30-plus foot of mass under the surface. Thiéry de Menonville additionally noted entrails or filaments extending out 80 feet like a mollusk and decided that based on the way the corpse undulated in the water, the skeleton had already dissolved. All in all, he thought it looked like a collapsed leather sack, with several splotches of black and brown.

Once the captain ascertained it was of no interest (i.e., had no commercial potential), the ship continued on. Finally arriving in Saint Domingue, Thiéry de Menonville established a plantation of the cactus in the Royal Garden at Port-au-Prince. He was rewarded with the title Royal Botanist, but his glory was short-lived. Within two years, he was dead of yellow fever. No one else had been instructed how to maintain the cacti, and as they quickly died, so did the cochineal insects, ending the experiment.

His notes were published posthumously and by 1812 has been translated into English by John Pinkerton, who, for lack of a better term, called the decomposing creature a leviathan. Yet, when England's Poet Laureate Robert Southey developed an interest in the increasing number of kraken reports, he was directed to Pinkerton's translation of Thiéry de Menonville. In an 1815 letter, Southey, a self-professed skeptic, accepts that the encounter at face value as being a kraken.

Suddenly, an account describing a decomposing mass, written in French, translated into English, and edited by several publishers, was a giant squid. It was written by a botanist with no formal training in marine biology and who was desperate to just get the ship back on course. In spite of this, Thiéry de Menonville's description does offer an identification.

It's the reference to the remains looking like a collapsed leather sack that provides the identification. The odoriferous discovery was a baleen whale in an advanced state of decomposition. Usually, when a whale dies, it sinks to the bottom of the ocean, where the corpse becomes an entire ecosystem of scavengers ranging from worms to sharks. However, if the whale received internal injuries in a collision with a ship, was poisoned by a red tide bloom (the earliest officially cited bloom was 1792 near Vera Cruz), or beached, the lack of external injuries would not allow the gases created during decomposition to escape. Instead, the animal would inflate like a balloon—stretching and distorting its external appearance. The whale would float instead of sinking, particularly in the warm Caribbean waters, which tend to fuel faster bacterial growth. The rapid gas production in the bloated whale carcass would eventually hit critical mass and the whale would either explode or rupture, collapsing like Thiéry de Menonville's "collapsed leather sack."

The distortion from the bloating would explain the botanist's questionable length of the carcass, which would place the corpse at the upper size range of the blue whale (*Balaenoptera musculus*). White foam is generated by decaying proteins collecting on the water surface and could have an oily sheen. The fact that Thiéry de Menonville did not note the water had also a reddish tint from a combination of blood and grease suggests the creature had died elsewhere and drifted. With the botanist's weather complaints (he keeps referring to the seasonal tropical squalls as hurricanes), the whale could have beached and washed back to sea, or drifted. This encounter appears to be a whale, which like Thiéry de Menonville, had a bad sense of timing.

## The Steamer *Neptune*

On January 3, 1830, the steamer *Neptune* left Matanzas sailing west along the northern coast of Cuba as she headed for Havana. At midday, the crew spotted a boat on a reef four miles away. Captain Jose-Maria Lopez ordered the ship to approach the vessel in case aid was needed. As they neared, they discovered the "boat" was actually a "fish of immense magnitude," rising 16 feet out of the water. It was surrounded by innumerable fish of different sizes, forming a mile-wide area around the beast. As they came close, the mouth opened and a terrible rumble sounded similar to that produced by a landslide. The noise convinced the crew and passengers it was time to resume their itinerary. It was noted there was a fin, black in color, and nearly nine feet high about 60 feet from his mouth. The tail was underwater, so the creature's full length could not be determined.

Captain Lopez reported the encounter to the harbormaster in Havana, more as a curiosity than a navigation concern. It was bigger than any whale he had encountered and shaped so differently that he assumed it was an entirely different species. Naturally, the local newspaper picked the story and it went global.

The first English language newspaper to translate the story was the *Charleston Courier*. That translation calls it a fish, and mistranslates "fin" as "wing." The French magazine *Revue Des Deux Mondes* correctly translated the article as a whale. That's the end of it for 75 years—a brief interesting report of an unknown whale.

Based on the size and dorsal fin, the *Neptune* encountered the bloated carcass of a Fin Whale (*Balaenoptera physalus*). Adult males can measure up to 78 feet and females are slightly larger. The reference to the mouth opening and a landslide-type noise being heard was gas being expelled. In the process of decomposition, methane and other gases build up in the whale carcass. The buildup of pressure could also distort the corpse's shape, rendering the species unfamiliar to don Jose-Maria Lopez.

In 1905, the *Neptune's* whale was resurrected by Comte Georges Gautron as proof of the existence of the sea serpent. Heuvelman reiterated it was a whale, specifically a beached whale rather than a floating gas-filled carcass. Heuvelmans specifically wondered why Gautron would include it in his book, and Heuvelmans has a valid point—Gautron's version is a badly garbled translation from the original Spanish; he should have referred to the perfectly usable version in *Revue Des Deux Mondes*.

Heuvelmans actively disliked Gautron's work, at one point referring to him as a "bearded dandy" in his *In the Wake of the Sea-Serpents*. Ironically, Heuvelman's decision to lambaste Gautron is the only thing that has kept his work from obscurity.

## The Barque *Ville de Rochefort*

On the 21st of April 1840, the barque *Ville de Rochefort* was in the Gulf of Mexico. Sailing from Vera Cruz to Havana, the weather was so good that Captain d'Abnour sent men aloft to unfurl the grand cacatois on the mainmast, only used when winds were light and favorable. One of the crew spotted seafoam on the horizon off the port bow, indicative of shoals or a reef. This was odd since the nearest land was the Yucatán Peninsula, 250 miles to the south.

After a few hours, the ship was close enough to see what appeared to be a long chain of rocks, gently rising at each end with the middle only a few feet over the level of the sea. The sea was breaking gently against these rocks, creating the foam. As the ship moved nearer, the captain noticed that the "rocks" were changed shape and position. It was obviously not a reef.

Captain d'Abnour called for his telescope. It appeared to be a long chain of enormous rings, resembling a series of barrels linked together, segmented like the back of a silkworm. As he continued to observe, he saw the far end was an enormous tail, longitudinally divided into two sections, white and black. This tail appeared to wind itself up and repose on a part of the thing itself. Then, at the other extreme, he saw a membrane rising to the height of about two meters (6 feet) from the water, and inclining itself at a considerable angle upon the mass. This led the Captain to conjecture that the monster had an apparatus for the purpose of respiration, similar to a lamprey eel.

As they passed by, they saw an antenna rising from the water to the great height of nearly eight meters (26 feet), terminated by a crescent of at least five meters (16 feet) from one extremity to the other. The ship continued on its way, not approaching close enough to identify the beast, but the Captain believed that it was an enormous serpent of at least 100 meters (328 feet) in length.

Upon return to France, the encounter was published in the *Journal du Havre*, a major French newspaper. The *Journal du Havre* was inclined to believe Captain d'Abnour. The editors didn't know marine biology, but they knew the captain—he was rebuilding his reputation after a streak of bad luck had embroiled him and the ship in a pair of unfortunate lawsuits. In September 1838, the mainmast had been struck by lightning in Havana and the insurance firms were refusing to pay for the replacement on a contract technicality. It was a nuisance at best, but the timing was terrible—it came on the voyage immediately after a January 1838 trip along the same route that also ended in a lawsuit. The *Ville de Rochefort* had encountered a tropical storm so severe the ship was at risk of foundering, and d'Abnour had jettisoned 30 barrels of tobacco powder to lighten the vessel ($400,000 worth in today's money). With the captain in court, the ship could not leave port, and an idle cargo vessel costs the owners money and the captain cachet. The last thing the captain needed was a frivolous sea serpent report when he was trying to keep a low profile and do his job without further incident.

The *Journal du Havre* article was immediately reprinted in other French papers and then translated into English by *The Standard* in London on September 15, 1840. From there, it spread across the British Isles and on to American newspapers as well. Oudemans and Heuvelmans each include Captain d'Abnour's account in their books.

Oudemans expresses some concerns about the report. He was convinced that Captain d'Abnour had seen a sea-serpent resting on the surface, showing numerous coils above the water but the head and tail being underwater and invisible. However, Oudemans was utterly at a loss to explain the various described features from diverse parts of the animal kingdom, such as a membrane, or what relevance it would have to an eel, let alone the antenna-mounted crescent.

Heuvelmans is less kind, sniggering that "Captain d'Abnour was clearly ambitious of writing scientific literature at its most pompous." He is correct in stating that a lamprey eel does not have a retractable membrane, suggesting perhaps it was a fin. Because of Heuvelmans's cavalier attitude, the case became an outlier and ignored.

What neither of these early cryptozoologists did was check the source material. Oudemans at least admits he never saw the original, using an English version that ran in *The Zoologist* in 1847. Heuvelmans cites the *Journal du Havre*, but leaves a glaring clue that he either liberally borrowed from Oudemans or also used the same source—*The Zoologist* ran the article with the incorrect date, citing the *Journal du Havre* article of September 15, 1840. That's the date the article first appeared in English in *The Standard*. This is significant as it means that *The Zoologist* also never checked its sources. The London newspaper's translation is wildly inaccurate and the source of all the confusion.

The original French newspaper article on August 26 tells a far different story. There is no "membrane." The word in question is *membrure*, a shipbuilding term. The English equivalent would be the ship's ribs, the transverse frames supporting the ship's planking. In other words, Captain d'Abnour did not see a membrane unfurling. He saw something rise six feet out of the water that reminded him of a ship's hull and the parallel lines of the ship's ribs. D'Abnour was not a biologist, so he attempted to explain biological features in familiar terms. Similarly, the word antenna does not appear in the article: *"...enfin nous vimes s'élever hors de l'eau une partie énorme de six à huit métres, terminée par un croissant ayant au moins cinq metres d'une extrémité à l'autre..."* (finally we saw a huge [body part] of six to eight meters rising out of the water, ending in a crescent of at least five meters from one end to the other).

So, instead of membranes and antennae, we're looking at a torso with parallel lines, rising six feet out of the water, then a tail of 19-to-26 feet, terminating a crescent of 16 feet. Captain d'Abnour is describing cetacean behaviors known as "spy-hopping" and "lob-tailing." Spy-hopping occurs when a whale holds itself vertically in the water and uses its tail fluke in order to maintain its head above the waterline. Lob-tailing describes a cetacean lifting its tail fluke out of the water to forcefully slap the water. And the size of the fluke or crescent means the captain has encountered a pod of humpback whales (*Megaptera novaeangliae*).

Assuming the "sea serpent" was a group of humpback whales, the remaining cryptic description begins to make sense. Humpbacks also have a varying number of bumps on their heads, called tubercles, each containing one hair that serves an unknown purpose but is assumed to collectively be a sensory array of some sort. To a non-biologist such as d'Abnour, the most comparative thing he could come up with to explain the tubercles along the whale's head was a lamprey eel's seven external gill slits on each side of its head. The ship ribs become the ventral grooves or pleats that expand during filter feeding.

Keeping in mind d'Abnour's observations were done at a distance with a telescope, the appearance of a long chain resembling a number of barrels linked together may be the captain mistaking six or seven whales for one 300-plus-foot serpent. The close proximity of the whales to each other and the reports of foam may indicate a unique feeding technique employed by humpback whales—bubble-net feeding. The whales swim in a circle around a school of fish, blowing bubbles in a circle and corralling the fish in an increasingly small area until all whales simultaneously feed on the trapped fish.

The *Ville de Rochefort* was not the only ship to encounter humpbacks. In 1841, American explorer John L. Stephens was in the Yucatán Channel on the becalmed Spanish brig *Alexandre*, attempting to travel from Sisal, Mexico, to Havana, Cuba. On the evening of July 14, he spotted "an enormous monster with a black head ten feet out of water" heading toward the ship. The captain, looking at it from the rigging with a telescope, insisted it was not a whale. A second one appeared near the craft, but regardless of the captain's declaration, Stephens was relieved to hear the whales spouting and seeing a column of water thrown into the air. As the sun set, they were lying motionless on the surface of the water, exhibiting another cetacean behavior known as "logging," resting without any forward movement at the surface with the dorsal fin and part of the back exposed; the logging behavior is displayed by other species of whales as well. Stephens' reference to the spout as a column identifies it as a humpback. The spout varies in height, shape, and visibility among species, and columnar is how a humpback's blow is described. The difference between Stephen's account and d'Abnour's is that Stephen's had no need for translation.

Captain d'Abnour did not see a giant sea serpent with a respiratory membrane and an antenna. A victim of a bad translation, he may have had a more significant chance at history—the captain's description, accurately translated, is the earliest published record of humpbacks bubble-netting until 1929.

## The Barque *St. Olaf*

The *St. Olaf* arrived in Galveston, Texas, on March 15, 1872. The ship was a regular at the docks; she averaged three trips each year from Liverpool, England via Newport, Wales. This trip was slightly different. Captain Hassel was new to the vessel, and there had been an encounter with a sea serpent.

Two days earlier, at latitude 26° 52', longitude -91° 20', the *St. Olaf* was 300 miles southeast of Galveston. Using the northerly Gulf currents to her advantage, the ship was two days out of port. Captain Hassel noticed a shoal of sharks passing the vessel. About two minutes later, one of the men reported that he saw something off the bow, like a cask on its end. Presently another crewman called out that he saw something rising out of the water like "a tall man." As the ship came closer, they discovered they were about 200 feet from a sea serpent, with its head out of the water. The serpent was unperturbed by the ship, lifting its head up, and moving its body in a serpentine manner. The crew could not determine the length, but what they could see was 70 feet long and uniformly six feet in diameter except for the head and neck, which were smaller. Hassel noted the head was flat and serpentine. The serpent had four fins on its back, and the body was greenish-yellow, mottled with brown and white underneath. The crew observed it 10 minutes before it moved away.

The report generated little interest. *The Galveston Daily News* mentioned it and ended the 150-word coverage by noting they thought the serpent was there to "swallow the wicked crew of Radical corruptionists who will soon be submerged in the sinking ship of State Government." That was the end of the matter as far as Texas was concerned. The *St. Olaf* unloaded, reloaded, and departed on June 6 for Wilmington, North Carolina, then Havana, Cuba, arriving back in England in August.

Texas newspapers may have been uninterested, but England was in the middle of a sea serpent craze, with so many sightings reported to the British papers that Heuvelmans would devote three chapters to discussing the accounts chronologically, and it is not a comprehensive list. J. M. Walthew, the shipping line owner and local agent in Galveston, was a major commercial agent in Liverpool before his immigration, and he was well aware of the potential free press for his Liverpool-Galveston shipping company. The Liverpudlian offices of Walthew & Co offices were run by his son J. Fred Walthew, who was advised to get the word out about Captain Hassel's sea serpent. Fred Walthew came through; *The Graphic*, a weekly newspaper, reprinted Captain Hassell's account as an affidavit, with Walthew listed as a witness to Hassell's signature. The article included two sketches by a crewman. The story was reprinted in British papers, but the *St. Olaf* encounter didn't catch the public's imagination and was soon was lost in the steady flow of sightings.

The sighting got a new life in 1892 with the publication of *The Great Sea-Serpent*. Oudemans quotes the entire article from *The Graphic,* before deciding it was either a hoax or an accurate account of what they saw, assuming that what they saw was four stragglers from the shark shoal, lined up perfectly with their backs and dorsal fins exposed. Oudemans bases his opinion on the sketches, as opposed to Hassell's text. The first sketch could be a line of sharks or dolphins, but the second image is decidedly serpentine. Oudemans dismisses that one as the crewman deciding in retrospect to construct what the serpent would have looked like on the surface. Oudemans doesn't address the non-shark coloration or the flattened head. More puzzling is why Oudemans thinks that four sharks can maintain a formation so well that it can fool a crew of seasoned mariners for 10 minutes.

Heuvelmans dismisses Oudemans's sharks swimming in formation theory, suggesting that a crew that had just seen a shoal of sharks pass by was not immediately going to mistake sharks for some other creature. He doesn't have any issue with the Oudemans claim of fraud per se, but question his motivation for the fraud claim. Heuvelmans explains that Oudemans was biased because he didn't believe in a sea creature with four dorsal fins. Heuvelmans was a strong advocate of a short-necked multi-finned marine cryptid, although later in the book wondered if the dorsal fins were actual lateral fins on a sea beast rolled over on its side. This is typical—Heuvelmans's book is based on assessing all sea serpent sightings and classifying them into one of nine categories that even Heuvelmans couldn't keep straight. The problem with his placing the *St. Olaf* serpent in the short-necked multi-finned marine category is that he specifically created it for armored whales, a cryptid that had already been debunked as fossils of two different animals jumbled together.

Because of the posturing by Oudemans and Heuvelmans, the *St. Olaf* encounter was mostly forgotten as a probable fraud. If mentioned at all, it was just one in a list of sea serpent sightings around the globe. Neither Oudemans nor Heuvelmans did extensive source verification (Ulrich Magin particularly takes issue with Heuvlmans's unchecked sources). As a result, *The Graphic* article is used as the prima facie source. They did not pursue the fact that a Norwegian vessel with a Norwegian captain might have further information in Norwegian media. Such an in-depth eyewitness account appeared in the Norwegian popular science magazine *Naturen* in 1893.

Headmaster Axel Conradin Ullmann was a teacher and a prolific naturalist, particularly in entomology. His collection of beetles remains a significant part of the University of Oslo's Natural History Museum. Ullmann was out exploring the fauna of Kragerø in 1886 when he happened into a conversation with another resident—Captain Andreas Hassel, now of the barque *Nora*. Ullmann recalled the encounter and, at Ullmann's request, wrote up the story again.

Ullmann passed the written report to Professor Robert Collett, curator of the Zoological Museum at the University of Oslo, for inclusion in the museum catalog of vertebrates. Collett summarily dismissed the report with a curt "First, Show me the beast!" Collett further antagonized Ullmann by admitting he barely glanced at the report, and it was probably just a giant squid or an octopus.

Hassel was adamant that it was not a squid or an octopus. He noted he had seen hundreds of octopuses. In fact, he recalled one time when his crew saw a large number of 6-to-9-inch long squid fly like a flying fish and fall down on the deck like a little rain squall.

The expanded version tells that the *St. Olaf* had just passed from the Yucatán Bank into deep water, the ocean depth dropping from 600 feet to over 13,000 feet. Sailor Jens Andersen was working in the rig when he became the first to see something. He sounded a warning—Second Officer Ole Thorbjørnsen reported that he saw something big rise up out of the sea in the distance. A few minutes later, Captain Hassel also saw something on the horizon that looked like a pod of cresting whales coming toward the ship. When they came closer, he saw that it was sharks. Hassel had seen and fished his share of sharks but had never seen or heard of such a large herd of sharks swimming at such high speed or breaking the surface like breaching whales.

Puzzled, Hassel talked to a fellow officer about this odd behavior for several minutes when again the warning was sounded—there were several big whales coming up on the starboard. The captain took the binoculars and saw multiple fins moving past each other. As the ship came closer, the crew saw it was one big animal with four fins on the back. When it came about 500 feet from the ship, it stopped, lifted its head six or seven feet above the water, and looked back and forth while the rest of the torso body tread water in a horizontal serpentine motion.

Hassel estimated the animal to be 130-to-160-feet long and about six feet in diameter in the middle. It reminded him of an eel in appearance but thinner at the neck and broader in the middle. The belly was white, otherwise it was the same greenish-yellow hue as a green moray eel, but with brown spots on the body. Green morays (*Gymnothorax funebris*) are one of the largest morays, but the eight-foot upper size range is a far cry from the length the captain saw.

The crew was nearing a panic, terrified that it was nothing less than the Great Norwegian *Soe Orm* described by Olaus Magnus. Captain Hassel knew it could not be the *soe orm*; although the length matched Olaus's Norwegian beast, the *soe orm*'s diameter was thicker (20-foot diameter) and it dwelled in subterranean caves along the coast. The *soe orm* also had a growth of hairs of two feet in length hanging from the neck, sharp scales of a dark brown color, and brilliant flaming eyes, all noticeably missing from the creature near the barque.

Hassel was concerned for other reasons. The ship was so heavily laden with railroad iron that she was low in the water—low enough that whatever the beast was, it could reach the deck.

The mate suggested firing a gun to scare it away. Hassel agreed, but by the time he returned with the rifle, the creature had taken off in the same direction as the sharks.

Upon arrival in Galveston, Second Officer Ole Thorbjørnsen made a drawing of the animal that was reproduced with the London newspaper account and has been extensively reprinted ever since.

Hassel had been a sailor for 33 years. He had seen whales and odd fish. He had encountered devilfish on the coast of Africa, and battled sunfish and swordfish—he knew the ocean, and he knew the animals suggested as candidates he could have mistaken for a sea serpent. The Captain was unswayed—the creature they encountered was something unidentified.

Ullmann contacted several other crew members who witnessed the encounter. Thorbjørnsen was on a voyage, but sailor Jens Andersen, steward John Hansen, and mate Knud Olsen all verified Hassel's account.

Ullmann was convinced it was a new species and had spent the previous five years trying to get a paper published in a science journal based on Hassel's report. But he was frustrated in the attempt because of Collett's stance. *Naturen* was a popular science journal, not a peer-reviewed journal, but it would have to suffice due to Collett's resistance. (Out of spite, Ullmann would donate his extensive collection of mounted insects to the Kristiansand Museum. They would not go to the University of Oslo until after Collett's death in 1913.)

Ullmann's interview with Captain Hassel did indeed add additional details, just not enough to suggest the encounter was with an identifiable creature. This means Oudemans and Heuvelmans, for lack of following up on research, may have missed a real cryptid, something that could not only force a shoal of sharks into a panicked retreat but also have no qualms in pursuing those sharks.

### The Steamship *Pecos*

The steamship *Pecos* left Galveston, Texas, and was heading to Miami for a quick freight delivery and then on to New York City with the rest of the cargo.

On the night of February 15, 1934, the freighter was making 15 knots, about 700 miles out of Galveston and about 100 miles south of Louisiana's South Pass and heading southeast, planning to reach the Florida Keys by morning. About 10 o'clock that night, Third Officer Sewell reported to Captain Leonard Baker that the *Pecos* had struck some sort of wreckage that was still hanging across the ship's stem. The captain went to the bow to investigate and could see there was something pushed along by the bow of the vessel.

Captain Baker ordered the ship to slow down to see if the debris would drop off without the water pushing against it, and then ordered the ship to come astern. By backing up, if the debris did drop off, he could make sure the ship got far enough away to maneuver past it safely without risk of fouling the propeller. As the ship moved backward, the "wreckage" began to thrash in the water—whatever it was, it was alive. Captain Baker and Mr. Sewell tried to identify it, but the night was dark and all they had was flashlights. The creature appeared to be 30-40 feet long. The exposed part above the water was five or six feet broad, and a dark grayish brown mottled color. It was too dark to see if the animal was injured by the encounter with the 4,572-ton steel vessel, but the *Pecos* was unscathed.

The lookout later reported to the Captain that there had been a jolt when the *Pecos* struck the beast, and Baker remembered feeling a shock at the time but thought it was the mate dropping the forward hatch, part of the nightly routine. The odds were not as good that the creature was similarly undamaged.

Captain Baker was a known and vocal skeptic as to the existence of sea serpents, but a collision in the middle of the Gulf of Mexico required a report to the Clyde-Mallory Line office upon arrival in New York on February 21; the captain mirthlessly filed an "unusual occurrence report" with Captain William Park, the line's general superintendent. In spite of carefully avoiding speculation other than it was a "sea animal," Captain Baker's timing was the problem. The Pecos encounter took place soon after the *Mauretania*'s first two sea serpent encounters in the Atlantic.

H. G. Wenzel, the passenger traffic manager, released Captain Baker's report to the press, hoping to divert a little of the *Mauretania*'s spike in bookings from the Cunard Line to Clyde-Mallory. The press obliged—and they also had a series of sightings at Loch Ness to utilize.

Captain Baker made his report and then stayed in his cabin until the ship was ready to depart. The home office delayed departure to allow an inspection of the bow to be sure there was no damage. Adding insult to injury, the delay meant the *Pecos* was leaving New York as the *Mauritania* was arriving.

The *Pecos* returned to Galveston a day late, making the schedule obsessed Captain Baker unhappy. When the Texas reporters cornered him, it did not improve his mood. He maintained his skepticism about sea serpents in general and Loch Ness in particular. He suspected that the *Pecos* had struck a whale shark (*Rhincodon typus*), the slow-moving filter-feeder. The average whale shark has the proper length and the "dark grayish brown mottled color" falls within the whale shark's colorization range. Baker did qualify the statement by noting he had never heard of whale shark colliding with a ship, which has become increasingly untrue as boat traffic and speed have increased since 1932. Baker also wished the collision had taken place in daylight so he could be sure. Fortunately, *Mauritania* had just reported its third sea serpent sighting, which meant Captain Baker could go back to his routine without further sensationalism.

## SS *Steel Inventor*

Timing was everything for the steamship *Steel Inventor* (the original owner, U.S. Steel Products, began the names all their ships with their product). On March 6, 1934, five days after the captain of the *Pecos* reiterated his disbelief in sea serpents, *Steel Inventor* arrived in New Orleans with a sea serpent sighting that was within 100 miles of the sighting by the *Pecos*. The *Steel Inventor* sighting, much like the *Pecos*, was overshadowed by *Mauritania*'s triple sighting. Unlike the *Pecos*'s Captain Baker, the witnesses aboard the *Steel Inventor* did not discourage the media.

The Isthmian Line freight ship had left Honolulu, passed through the Panama Canal, and was steaming to New Orleans, the first stop in what was normally a series of short stops along the coast to its homeport in New York. This trip was different; *Steel Inventor* was staying in New Orleans for scheduled maintenance. The voyage had started back in September in New York and visited England, Ireland, and Canada before refueling in Key West and on to Hawaii. *Steel Inventor* was a rugged ship, designed for these extended cargo runs, but in 1921, the ship had collided with the American destroyer *Woolsey* of the Pacific Fleet off Coiba Island. The steamship cut the warship in two with minimal damage to *Steel Inventor*, except an ongoing issue in rough seas when the ship's heaving and pitching tended to loosen rivets on the steel plates more frequently than the norm. As a result, *Steel Inventor* was sent in for maintenance more often than her sister ships. This maintenance also gave the local newspapers time to chat with the sea serpent witnesses for local editions, with details that did not reach the AP newswire.

On March 5, Third Mate W. D. Day was on the bridge for the 4-to-8 a.m. watch with seamen Ralph Jones and Raymond Cousins. At three bells (5:30 a.m.), the ship was 250 miles south of South Pass when the sea serpent was spotted 100 feet off the starboard bow. Day went to notify the captain. Jones took the wheel and sent Cousins to fetch his camera.

Captain Boase arrived in a hurry, fearing it was a submarine. The 64-year-old Boase had spent over 15 years in the British Royal Navy and retired as a decorated commander. He had seen what German U-boats could do during the First World War, so his concern was justified. Pearl Harbor was seven years in the future, but sub activity had been reported along the increasingly nervous American coastlines. He was relieved to see it was only a sea serpent.

As the ship neared, the serpent disappeared. The men agreed the beast had a serpent-like head and the body was three feet in diameter with disc-shaped scales six-inches wide. They were unable to determine the serpent's length because only a few of its coils were visible at a time. The witnesses were all quick to emphasize there were no horns or whiskers, a direct reference to *The New York Times* attempt at illustrating the *Mauretania*'s sea monster.

A week later, reporter William Fitzpatrick strolled down to the docks one evening, where Boatswain Morgan A. Rock and Second Mate Winslow Monroe Rouse were supervising the monotonous task of stripping the old paint off *Steel Inventor*. Neither had seen the sea serpent and Mr. Rouse, after 11 years at sea, hadn't seen anything out of the ordinary. But well aware that his ranking officers had, he was careful to be noncommittal as to their existence.

Boson Rock, with 28 years at sea, had seen a lot of things, some even odder than sea serpents. He claimed he had seen the serpent that the captain had seen, but like most sailors, kept his mouth shut since no one would believe him. He claimed to have seen another one after the war, aboard the United States Army transport ship *Buford* as it transported casualties home from France, and another one in 1922 aboard the *Star of Bristol*, a two-masted schooner, in the Straits of Queen Charlotte Sound. Interestingly enough, Rock's 1922 sighting described the beast as having a horse-like head and long neck, identical to reports of the Cadborosaurus, reported in Cadboro Bay on the opposite side of Vancouver Island.

Perhaps the most interesting thing about the Steel Inventor is that the crew not only appears to have confirmed the steamship Pecos encounter, they nonchalantly offered an uncatalogued report of the Cadborosaurus that predates a major spike in sightings in the 1930s. Not bad for a ship with loose rivets.

# CHAPTER 4
## Caribbean Cryptids

**HMS** *Orontes*

The HMS *Orontes* was a steam-powered Royal Navy troop-ship assigned to transport troops to southern Africa and the West Indies. The *Orontes* was in the Mona Passage, the strait that separates Santo Domingo and Puerto Rico, bound from Jamaica to the Royal Navy based at Queenstown (now Cobh) in Ireland. At 3:15 a.m. on March 20, 1873, the ship developed problems with her air pump. It was not serious but a nuisance that required shutting down the engines and letting the ship drift in the gentle currents while repairs were made.

Four hours later, as Captain John Perry and the navigator on duty, Sub-Lieutenant Reginald Yonge, were walking on the upper bridge, Yonge spotted something white slowly rising out of the water in the nautical dawn light. Any growing irritation over the duration of the repairs vanished immediately.

It was "the head of an immense monster," which Yonge would later compare to the head of an eel. It rose out of the water and remained stationary. To the horror of both men, the *Orontes* was drifting right at it. Yonge ran to the chart house for a rifle, and Captain Perry could only watch as the head gradually sank and then rose again closer to the ship—almost so close as to touch it. The sea monster was apparently convinced the vessel was neither edible nor a threat. It turned slowly while submerging and swam away in a southwesterly direction. It surfaced several times as it departed, and each time the captain fired at it. The captain would later find that the quartermaster and signalman had witnessed the encounter from the lower bridge, and First Lieutenant William Lang, who had been awoken by a snorting nose, saw the creature cabin porthole. All the witnesses agreed it was 40 feet in length. The back of the head was black, the throat and belly were white, and the eyes were also white and set well back.

In another hour, the engines were fired up and the *Orontes* was again underway. The sea monster was not seen again. Yonge recorded the incident in his journal, and aside from being interviewed about the matter by Admiral Sir

Rodney Mundy, Commander-in-Chief at the Portsmouth Station when the ship arrived in April, the matter was forgotten. Yonge continued his military career before retiring as a commander.

Then, in late June 1906, Yonge was reading the *Illustrated London News* when an article by zoologist W. P. Pycraft caught his eye. The piece by Pycraft was about a recent sighting of a sea serpent by the yacht *Valhalla* on a research expedition off Brazil. This new sighting by a vessel with naturalists as witnesses appeared to validate the still debated 1848 serpent reported by HMS *Dædalus,* a frigate that encountered what its captain, Peter M'Quhae, described a 60-foot long serpent, 300 miles off the coast of southwestern Africa.

Yonge then recalled his encounter in the Mona Passage 33 years earlier and copied his journal entry into a letter to the *Illustrated London News,* which was published as a follow-up to the Pycraft article.

To emphasize the veracity of the story, he noted that Captain Perry was a now-retired Rear Admiral, and Lieutenant Lang was now a retired Vice-Admiral. Yonge closed his letter by clarifying several points from his journal, such as how silently the ship had been drifting. He felt the only reason they had gotten so close to the sea monster was that the silent vessel had accidentally snuck up on the animal, who he believed had been asleep.

In addition to her military service and sea serpent encounter, the HMS *Orontes* also has a notable literary credential. According to Arthur Conan Doyle, in his 1887 novel *A Study in Scarlet,* the *Orontes* is the troopship that brought John Watson back to Britain during his convalescence after the 1880 Battle of Maiwand, where he would become the roommate of Sherlock Holmes.

## The Adventures of the RMS *Mauretania*

By 1934, ship crews were paying more attention to the sea surface as they traveled, even if they weren't consciously aware of it. It was becoming evident that another war was looming. Submarines were a growing concern—the London Naval Treaty would be signed in April, a specific (and futile) attempt to regulate submarine warfare. But this added scrutiny of the seas had a beneficial side effect for the field of cryptozoology. When something broke the surface, you paid attention. As a result, reports of sea serpent sightings were coming in from across the globe at a rapid pace.

And no ship was paying closer attention to the seas than the RMS *Mauretania*—her sister ship *Lusitania* had been torpedoed during World War I killing 1,198 passengers and crew. The *Mauretania* was a Cunard passenger liner, the "Grand Old Lady of the Atlantic." It was launched in 1906 and held the transatlantic speed record for years. By 1934, her glory days were past. The aging ocean liner had just been relegated to Caribbean cruises but was still a favorite among her passengers.

On January 30, the *Mauretania* was steaming at 24 knots, about a mile off the island of St. Eustatius. S. W. Moughtin, Senior First Officer, was on the bridge when he spotted something on the starboard bow coming from the direction of the Saba Bank. He called Senior Third Officer J. W. Caunce over to confirm what he was seeing—a 65-foot long, shiny jet-black sea monster. The head was about six feet out of the water and was two feet across. About 45 feet of the body, which had a six-foot width, could be seen in serpentine curves on the surface of the sea, and based on the water disturbance, the seamen believed the body extended another 20 feet below the surface.

Being an efficient officer, Moughtin noted the encounter in the ship's logbook: "sighted sea monster headed S. W. 1:20 P. M." and drew a quick sketch of the creature. Moughtin and Caunce were mocked by fellow officers about the log entry and decided to keep the story to themselves in the future. Unfortunately for both, that decision was doomed.

On February 2, the *Mauretania* had anchored in the harbor of La Guaira, Venezuela. This time it was Caunce's turn to call Moughtin to the port side. It was an entirely different type of sea monster, barely 700 feet off the bow. This beast was 25 feet long and 15 feet across the back with two huge fins that stuck out six feet on each side. Moughtin's impression was that it was some sort of a cross between a giant ray and a skate. One fin would submerge while the other rose in the air, alternating as it swam past at two knots. The mouth was white and about three feet wide. This sighting was most likely indeed a giant manta ray feeding on the surface. That story never gained the traction of the first serpentine sighting.

Moughtin did not enter it into the logbook this time, but the passengers had seen the creature. When they disembarked in New York City on February 9, it didn't take long for word to get out. By Saturday, T. Walter Williams of *The New York Times* had tracked down Moughtin for an interview. He obliged, reluctantly. Williams was the *Times* ship news editor and "Skipper" Williams was legendary for his encyclopedic knowledge of the ships that frequented New York and his relentless coverage of maritime news. The Skipper was going to get his story, with or without Moughtin—Williams was also known to invent a story or two on a slow news day. If Moughtin wanted to come out of the story looking reasonably sane, he had to talk to Skipper Williams and give him the story before Williams could elaborate on it.

The *Mauretania's* Captain, Reginald V. Peel, confirmed the reliability of his officers, stating there was no doubt that a sea serpent had been sighted in the Caribbean by the two officers on the bridge, and that another had been seen in Venezuela. If Moughtin thought it was mortifying to be mocked by his fellow officers, he had to have been horrified by the article in the Sunday *New York Times*. The publicity was bad enough, with the comparisons made to other reports across the globe, including the latest tales out of Loch Ness. The story itself was reported in a straightforward manner, but the article included an illustration above the headline. The drawing looked more like a Chinese dragon than the sea serpent Moughtin described. Even worse, the wording suggested it was Moughtin's logbook sketch!

When the *Mauretania* docked in New York after its next trip on February 23, Williams was waiting for Moughtin again. Moughtin was not happy. His St. Eustatius sea serpent sighting had gone international, and the politest thing he could say was that he was surprised at the wide interest. He reaffirmed the description of the creature and, more tellingly, admitted if he ever saw another sea monster, he wouldn't tell anyone, even his own brother. He further emphasized that whoever drew the serpent from the previous article, it certainly wasn't him.

Skipper Williams also interviewed Sir Edgar Britten, captain of the RMS *Berengaria*, the Cunard line vessel that had taken the *Mauretania's* transatlantic route. Britten said that Moughtin had served with him on the Cunard line's RMS *Laconia* for a long time and had proven himself a very reliable officer. Britten ended the interview echoing what the *Mauretania's* captain had said— if Moughtin told the story, then it was true.

But the story Williams published in the *Times* contradicted the interview Moughtin gave to the *New York Sun*. In the *Sun* article, which was much less serious in tone, Moughtin admitted to lying awake rethinking the sighting, suggesting Moughtin was having nightmares. *The Sun* makes light of Moughtin's attempt to reposition his sighting as a misidentified pod of dolphins, cavorting in the waves single-file, but astutely notes that whatever the sighting was, Moughtin was getting fed up with the notoriety.

Moughtin had every reason to be concerned over about the continued coverage of the sea serpent—he was Senior First Officer, a senior bridge official responsible for safety and training of all ship personnel. It was just a step below Staff Captain, second in command on a ship. An embarrassment could jeopardize his career path—there was no shortage of mariners from the Great

War competing for the bridge positions.

Instead of quietly going away, things got worse.

At 1 p.m. on March 6, the *Mauretania* was 60 miles east on Nassau in the Bahamas. The ship was steaming at 26 knots, heading west to go around the island to anchor in Nassau. To his dismay, a mile and half off the port bow, heading east, Moughtin spotted a 60-foot sea serpent. It was the same officers, Moughtin and Caunce, on the same ship during the same bridge watch. This time, Moughtin called for the Captain. Captain Reginald V. Peel arrived in the bridge in time to get a good look at the creature. The captain, aware of the grief Moughtin and Caunce were already getting, decided that another seas serpent sighting was the last thing either needed. The deck officers agreed that no one would mention this sighting. Passengers again thwarted the plan to ignore the incident. Mrs. Robbins Ibeson of Chicago noticed the serpent, calling it to the attention to Staff Captain A. T. Mott, the second in command of the *Mauretania*. She also brought the serpent to the attention of Skipper Williams of *The New York Times* upon disembarking.

With Moughtin and Caunce avoiding the press, Williams went to the top and spoke to both Captain Peel and Staff Captain Mott. Peel admitted he had gotten a good look at the creature, nearly three minutes, and that it matched the description of the first sighting off St. Eustatius—jet black, 60 feet long, with four humps and four of what he assumed were fins. But he had no interest in being "mixed up in the sea serpent story business."

Mott provided more information, observing it was "over sixty feet long with undulating humps and a large head lying flat on the surface of the water." Mott, regarded as somewhat of an expert on whales along the North Atlantic routes, was convinced that the serpent was the same one sighted two cruises ago by Moughtin and Caunce. Mott ended his interview with an enigmatic statement—"I do not say that it was extraordinary, but it was strange—very strange. I believe there are many much bigger monsters in the sea, as I have seen some in my sailing-ship days in the Indian Ocean."

By all rights, this should have been the end of the matter, but on March 23, Skipper Williams was surprised by the arrival of Moughtin and Caunce at his desk. They had been puzzled by the version told by Captains Peel and Mott. Moughtin and Caunce explained that they had seen two sea serpents that trip. Peel and Mott had both seen a sea serpent, but unbeknownst to either, they had seen different sea serpents. The one that Peel had described was some distance away and had been viewed through binoculars. The one spotted by Mott and the passengers had taken place 15 minutes later and was so close it was "almost right under the bow of the ship." Caunce had spotted it, and Moughtin initially thought it was a fishing net. He adjusted course to starboard to avoid it, but the beast simply submerged before the boat could pass.

This would be the last sighting of any monsters from the *Mauretania*. The next month, the Cunard Line would merge with the White Star Line. The new Cunard White Star Line found it had too many ships and not enough passengers. The older ships were going to have to go. The line withdrew *Mauretania* from service in September 1934, accompanied by a flurry of protest letters from her many loyal passengers, including President Franklin D. Roosevelt. Her furnishings were auctioned off, and the *Mauretania* was sent to the scrap yards in Scotland.

If Moughtin was concerned with his career, his worry was for naught. As the *Mauretania* ceased operations, he was promoted Staff Captain of the RMS *Carinthia*. Moughtin developed an interesting defense mechanism to deflect his part on the sea serpent encounters—he camouflaged them by becoming known for his tall tales and fish stories. When he made his first trip to New York with the *Carinthia*, he met up with Skipper Williams again. This time the two exchanged yarns, and Moughtin's story of a giant shark who fatally swallowed a barrel of oil and whose eyes were tear-streaked from the pain appeared in *The New York Times*. But what could have been the start of a long series of tall tales between old salts was cut short when Moughtin received another promotion, Staff Captain of the RMS *Aquitania* cruising the Mediterranean. The *Aquitania* was the last surviving of Cunard's "grand trio," along with her sister ships the *Mauretania* and *Lusitania*. Although now on the fast track to his own ship, his career was cut short by fatal complications after emergency surgery for gastric ulcers in 1939. Stanley William "Pat" Moughtin (1884-1939) was buried near his home on the Isle of Man.

Neither did the cryptid encounters affect Caunce's career. He continued his association with the Cunard White Star Line. When he retired in 1962 after 42 years with the line, he was Commodore of the Cunard Line, captaining the flagship RMS *Queen Elizabeth*. He had achieved the rank of Commodore in the Royal Naval Reserve. He continued to work occasionally with the Royal Navy on panels investigating naval wrecks and collisions. John William Caunce OBE (1896-1969) was buried in his native Lancashire.

The new RMS *Mauretania* was launched in 1938. Like her predecessor, she cruised the Atlantic, was converted to transport during a war, and by 1964 was obsolete and used for Caribbean cruises from New York before she too was scrapped in October 1965 in the West Indies. She was never as popular as the original *Mauretania*, and more importantly, never saw a sea serpent (or at least never reported one).

## The Nun's Sea Serpent

Sunday morning, Sept 25, 1960, taxi driver Mansfield Robinson picked up his regular fare in Hamilton, Bermuda. Three nuns were making their weekly trip to St. Anthony's Church in Warwick, about 10 miles away. It was a trip they had been making for a decade, and the route rarely varied, following Pomander Road to Harbour Road along the eastern edge of Hamilton Harbour, mostly for the scenic view. Robinson always slowed the car to a leisurely 15 mph so his passengers could enjoy the view.

That morning, one of the women, Sister Jean de Chantal Kennedy, observed that a small inlet called Red Hole near the north end of Harbour Road looked particularly picturesque, with no larger ships to mar the view of the small boats at their moorings framed by Inner Harbour in the distance. She noticed a long gentle wave, like the wake of a passing motorboat—except no boats were moving, motorized or other. One of the passengers thought she saw divers surfacing, although Sister Jean thought it was a series of "black blobs." The taxi passed a copse of oleander obstructing the view. As the harbor came back into view, Sister Jean saw the black blobs now looked more like partially submerged tires, being towed quickly by a motorboat—except there were still no boats moving. She wondered out loud if they were seeing a sea serpent, which one of her fellow passengers admitted was her thought as well.

As the road became more rural, they could follow the serpent whose direction and speed was roughly the same as the taxi. When the snake slowed down, Robinson stopped the vehicle near the Lower Ferry. The driver and passengers could examine what they now definitely considered a sea serpent. Less than a quarter-mile away from the beast, they noted a disturbance in the water behind the humps, which they presumed was the thrashing of a tail underwater. Then a serpent's head rose in the front to about the height of the first hump. Sister Jean estimated that the beast was 25 feet long, possibly 30 allowing for the unseen tail.

It appeared to scan the shores, then submerge its head and neck again, with a motion the sister described as swan-like. The skin was shiny black and smooth and appeared as it would be hard to the touch. The coils were about two feet high and arch-like, clearing the water—they could see beneath them. The coils did not undulate, or appear to be involved in moving the serpent forward. The head was away from the group so they could not see the face, but there was no change in circumference between the head, neck, or body. It gently sank from view, with the tops of its coils barely above the waterline, all but unnoticeable if you did know where to look. There was something else on the surface: a dorsal fin. Since they did not see a fin on the beast, they assumed it was prey.

Kennedy made a few calls, first to the acclaimed diver and explorer Teddy Tucker. Tucker, a regular advisor and guide for marine biologists and underwater photographers, believed her. He recalled a black snake-like carcass that had washed ashore two years before at Cambridge, less than five miles straight across Great Sound from Red Hole. It was too badly decomposed to identify, and it was only eight feet long, but the skin was also hard.

Her second call was to naturalist Louis S. Mowbray, curator of the Bermuda Aquarium, who would normally consider a Sunday morning sea monster as a lingering effect of a Saturday spree. However, he was hard-pressed to dispute three nuns, especially when one was Sister Jean, an instructor at Mount St. Agnes, an award-winning local historian, and a longtime member of the National Trust.

That afternoon two more sightings were reported, and whatever misgivings the aquarium had were overridden as the press caught wind of the story. The aquarium boat went out Monday morning to investigate a whale sighting, and then to Red Hole to look for "Sister Jean's Sea Serpent." They found nothing, but by the time the ship departed and returned to the aquarium that afternoon, another sea serpent report had come in from Red Hole.

There were more sightings, as people flocked to the shore to look for the beast. Many reports came from Red Hole and the Inner Harbour, but there were sightings in the deep sea and around nearby small islands. The eyewitness reports were remarkably consistent except for one sighting in September when an Air Force civilian employee and his family saw the serpent following a cruise ship out of the harbor. He estimated the length as being closer to 100 feet. The last sighting was on October 12. The serpent was spotted off the North Shore, heading toward North Rock. The locals joked that the serpentine tourist had finally headed home.

When Sister Jean wrote of the serpent in one of her later books, she admitted the finally understood why so few sea serpent encounters were actually reported—15 years later, she was still annoyed by the reporters who questioned her sanity, sobriety, and eyesight. But she did feel like she got the last laugh on the naysayers—a calypso song was written about the whole affair. "The Nun's Sea Serpent" ended with the lyrics assuring the listeners the story was all true!

## The Haitian Water-horse

As 1961 began in Haiti, sporadic reports of a sea serpent sighting started to filter in from the beaches of Côte-Plage. It had garnered some press but was far enough away from the tourist areas to remain a local matter. Descriptions varied on the length, but all agreed it was big, black in color, and had a head resembling a horse.

By February, the novelty had worn off. A young boy had been attacked by a shark in local water and lost an arm. Blame quickly evolved from a shark to the sea serpent. So, when the creature made another appearance the last week of February, it made the inopportune choice to surface near the docks at Bizoton.

The Bizoton Docks were also used by the neighboring Haitian Coast Guard station, a hybrid Navy/Police force that immediately dispatched cutters to give chase. The Coast Guard lobbed shells at the cryptid, which was hit by several rounds from machine guns. Witnesses reported the sea serpent made a mournful cry that sounded almost human. Other water-horse accounts claim merhorse-type sea serpents have almost vertical diving capabilities, meaning it could have survived the onslaught. Although there is no confirmation as to whether it escaped or was killed, no further sightings were reported.

The brief encounter remains of interest because it is one of the few water-horses reported in the region, and because, even more notably, it was a rare account of vocalization by a marine cryptid.

## The *Alvin* Plesiosaur

*Alvin* (DSV-2) is a deep-ocean research submersible owned by the US Navy and operated by the Woods Hole Oceanographic Institution. An invaluable resource in deep-sea research, it is known to the public for locating a 1.45-megaton hydrogen bomb lost off Spain in 1966 and for the exploration of the wreckage of RMS *Titanic* in 1986.

The *Alvin's* first deep-sea tests took place off Andros Island in the Bahamas in 1965. On 20 July, *Alvin* made its first 6,000-foot manned dive with pilots William O. Rainnie and Marvin McCamis. The dive was twofold in purpose. Officially, it was for US Navy certification. Off the record, it was there to inspect the cables of the Artemis Project, a top-secret underwater listening array to detect soviet subs approaching the US coast.

According to Charles Berlitz of Bermuda Triangle fame, Rainnie and McCamis saw a plesiosaur on the dive. The *Alvin* was in a crevasse below the 5,000-foot mark following the cable when McCamis noticed motion. After confirming the sub was not drifting, he thought it might be a utility pole from the original installation. He swung *Alvin* around only to see "a thick body with flippers, a long neck, a snakelike head with two eyes looking right at us. It looked like a big lizard with flippers—it had two sets of them." It turned and swam away before the cameras could be adjusted. McCamis entered it in the dive log, but the Navy did not include it in the published record.

The only source for this cryptid encounter wrapped in government cover-up is a book of questionable merit, Charles Berlitz's *Without a Trace*, a book rumored to have been rushed into production to refute Lawrence David Kusche's book debunking Berlitz's previous title. The text is poorly edited—McCamis is not identified by name until 440 words into a 500-word discussion, most of which was a Q&A with McCamis but still gets the year wrong. Also questionable is the credibility of Berlitz, which Kusche suggests is "so low that it is virtually nonexistent. If Berlitz were to report that a boat were red, the chance of it being some other color is almost a certainty."

A plesiosaur is unlikely for several reasons. A mile below the surface, where the water temperature is about 40ºF, the reptile would need to be homeothermic, meaning that it was able to regulate its body temperature, a relatively new and still conceptual theory. And for the air-breathing reptile to go that deep, their chest structure would have to have evolved to resemble an architecture more similar to a marine mammal, with lungs able to partially collapse, allowing oxygen and carbon dioxide to be absorbed by the animal's bloodstream, but not the nitrogen that would lead to the bends. So radical physiological evolution aside, a marine reptile designed for shallow water (i.e., well-lit) feeding, would have no reason to be so deep where prey was all but invisible.

An alternative might be the deepsea lizardfish (*Bathysaurus ferox*). Although recorded adults barely reach two-and-a-half feet in length, the species is rare and could grow larger in the inaccessible depths. Additionally, we have no indication of the size of the creature McCamis and Rainnie think they saw, and we have no way to calculate magnification through the porthole or any possible memory distortion caused by being in a high-stress environment and having something coming toward the *Alvin* out of the darkness, barely illuminated in the exterior spotlight. It would not take much for a slender, cylindrical body and a lizard-like bony head with an enormous mouth filled with needle-like teeth coming toward the sub to be imagined as the head and neck of a plesiosaur. The eyes the McCamis noted match the eyes of a lizardfish's, which are pronounced with large pupils. Sunlight does not reach the environment the *Alvin* was in, but the deep-sea lizardfish is an apex predator whose eyes can detect the bioluminescence of potential prey, or the lights of a deep-sea sub.

Berlitz would mention the encounter again briefly in his 1981 on the impending end of the world in 1999; he saw it as proof that dinosaurs like the *Alvin* plesiosaur and the Loch Ness monster survived the cosmic irradiation of the previous apocalypse by going underwater. The only non-Berlitz published record of the *Alvin* encountering a sea monster is an account of a shallow (120-150 feet) dive off Norfolk, Virginia, when two passenger biologists dubbed a 25-foot-long chain of unusually large salps (*salpa vagina*) they saw as a "sea monster."

## The Bermuda Sphere

In November 1969, Richard Winer and Pat Boatwright were diving at 35 feet in waters that plunged to 1400-foot depths, 14 miles southwest of Bermuda. The two were photographing the underwater movement of a telemetry buoy's moorings in rough seas for General Electric. Winer had used all the film in his camera and was preparing to head back to the surface when Boatwright gestured for him to look down.

Both men were former Navy men and experienced divers, but neither recognized the object that was slowly rising beneath them at a depth of 100-150 feet. Winer estimated the size of the object was 50-to-100 feet in diameter and nearly perfectly round. And whatever it was, it was alive. Its color was a deep purple with an outer perimeter that seemed to be pulsing, but they could not observe any movement of water that would suggest the pulsation was locomotion. As Winer and Boatwright started for the surface, it stopped and slowly descended and vanished into depths.

Five years later, the encounter still perplexed Winer sufficiently to include it in his book on the Bermuda Triangle. In spite of the publisher's breathless hyperbole on the cover, Winer believed that unpredictable weather and poor seamanship could explain all the mysterious phenomena in the area. The book notes that Boatwright had decided it was a giant squid, but Winer felt the movement of the creature was more similar to that of a jellyfish. Cryptozoologist Gary Mangiacopra concurred with Winer but noted the largest recorded size for a jellyfish was a lion's mane jellyfish (*Cyanea capillata*), a species that lives in cold water. That specimen was measured in 1865 at seven-and-a-half feet wide, a far cry from Winer's 50 foot or more diameter.

Mangiacopra concluded that this is unquestionably proof of a giant jellyfish living in the deep waters of the Caribbean. Neither he nor Winer had another alternative, but there is one: in the 1970s, when Winer and Magiaco-pra were corresponding, there was another choice, one that would have been virtually unknown outside marine biology circles—squid egg masses.

A squid egg mass is a gelatinous substance that expands to enormous sizes. Danna Staaf, a squid expert who studied a Humboldt squid (*Dosidicus gigas*) egg mass in the Gulf of California, described it as between 10-and-13-feet long. And although the Humboldt squid is limited to the Pacific, the neon flying squid (*Ommastrephes bartramii*) is found across the globe, including off Bermuda. In 2015, a neon flying squid egg mass was encountered by divers off Turkey: the egg mass was a sphere measuring 13 feet across and was filmed by *National Geographic*. This is smaller than Winer's estimated size, but he was unsure of distance and size, and does not state whether he was allowing for the refraction of light through his diving mask, which makes objects appear about 34 percent bigger and 25 percent nearer when underwater.

These masses are rarely seen because they are usually laid too far offshore, too deep for divers to encounter, and only take three days to hatch. But as Sta-af notes, they do occasionally drift to shallow water. Winer specifically noted he was diving in rough seas, which would be perfect for pushing a gelatinous ball of mucus off course before it again begins to sink to about 500 feet. That depth is the pycnocline, or the boundary where the density changes from sur-face waters to deep seawater, and it's where the eggs achieve buoyancy at a depth optimal for the squid to be born.

In this scenario, two divers exhausted from battling the elements spot an errant neon flying squid egg mass as it appears to drift upward. It seems to pulsate from the buffeting current. The female may have recently extruded the egg mass at the depth where the divers encountered it, the density of the mass still in flux as it absorbed seawater before beginning to descend slowly.

The egg sacs are translucent, but Winer notes the encounter happened in the late afternoon with light coming at distorted angles, so it is possible an-other spectacular Bermuda sunset was beginning, sending vivid colors at the distorted angles and reflecting off the surface. At 30 feet down, the little bit of red light making it to that depth would be reflecting purple.

So several circumstances have to take place for this to be a squid egg mass. But since Winer's entire book is based on the premise that the Bermuda Tri-angle is explainable natural phenomena, it would undoubtedly be within the realm of possibility.

**The Monster of San Miguel Lagoon**

In August 1971, a rumor circulated through the city of Havana about a horned monster rising out the water in a flooded quarry at San Miguel del Padrón known locally as the San Miguel Lagoon. Word spread and local curiosity was piqued, so much so that the Havana suburb was seeing increasing crowds hoping to catch sight of the monster.

The story quickly evolved from cryptozoology to folklore. It was said the monster appeared to an old man who lived on the lagoon, driving him suicidally insane. As historian and author Mario Masvidal Saavedra later discovered, there had been no newspaper reports on the cryptid. Nonetheless, thousands were flocking to San Miguel Lagoon through word of mouth.

Havana's Radio Progreso featured a large segment on the monster, featuring reporters who interviewed onlookers, many of whom claimed to have seen the beast. But no one agreed on the description, which ranged from a black ball that resembled a horned hippopotamus to an eyeless spindle-shaped creature. Then one of the reporters saw something with a rounded shape and a rough texture rise from the green waters, then sink again.

Radio Progreso's signal reached Miami's Cuban expatriate community, and soon after the UPI wire service ran the story. The coverage was nationwide, but the radio program's broadcast was still overlooked by most of the newspapers. A local charcoal maker, who had worked near the lagoon for a decade, identified the monster as one or more palm tree trunks being lifted to the surface by a buildup of gas as the garbage dumped in the lagoon decomposed. He recalled seeing this happen several months before when a fleet of trucks pulled up and dropped load after load of tree trunks into the water.

Since Radio Progreso was (and is) a government-controlled station, the Monster of San Miguel Lagoon was officially identified and debunked. In the United States, the story lingered until the end of 1971 as a filler article, giving the monster a longer life outside of Cuba than in its own lagoon.

**The St. Lucia Thing**

Anse Chastanet is a boutique hotel near Soufrière on St. Lucia's southwestern coast. It is in the heart of St. Lucia's marine reserves. One of its amenities is Scuba St. Lucia, founded in 1981. The resort is on the shore of a pristine coral reef, establishing Scuba St. Lucia as one of the world's top diving destinations.

From the beginning, the night diving tourists reported seeing a worm-like creature that was described as 10-to-15-feet long and as thick as a human arm. The worm was extremely sensitive to lights; any dive light was enough to make it disappear into the coral. Some divers were sufficiently startled by their encounter with a massive worm that they reported it to the local marine biologist. The creature was so incredible, it was considered more folklore than zoology and was simply known as "The Thing."

In 1994, Walt Stearns, a professional marine photographer, heard of the creature and decided to look for it himself. Stearns explored the reef at depths between 40 and 80 feet, keeping his hand cupped over his dive light to minimize the amount of light reaching the reef. After 45 minutes, he spotted movement. Compared to what the tales described, this worm was comparatively small, only about four feet in length. The worm was divided into round segments, with pronounced bristly "feet." The segments were a reddish copper brown, with iridescent speckles of bright reds, yellows, oranges, blues, and greens. Contrary to the stories, the worm was unbothered by the camera's strobe light, which Stearns hypothesized indicated an ill or injured specimen. Stearns had never seen a worm of this size, let alone one 6-to-15-feet long, as claimed by some witnesses. He sent photos out for identification, to no avail.

The general consensus is that this was a new species with features typical of the marine worms known as *polychaetes*. Polychaetes can grow to six feet or more, and there are several giant polychaetes in the Caribbean, but generally the majority of species are under a foot in length. All are segmented and characterized by hair-like structures called setae. A pair of these setae are mounted on appendages called parapodia, which can run the length of the body.

Joan Marsden, a marine invertebrate researcher and polychaete expert at Montreal's McGill University, believed it was a member of the *Eunicidae* family of polychaetes. Marine biologist Ben S. Roesch concurs, noting the similarity to the Red-Gilled Marphysa (*Marphysa sanguinea*), which is also iridescent. That species also lives in tunnels and is native to the West Indies. The Marphysa only grows to lengths of two feet, large for a polychaete, but not even close The Thing's length. Marsden noted that the difference between the head characteristics of the various species is the most definite way to pinpoint a polychaete, and she had yet to see the head of a specimen of The Thing. Much like the Red-Gilled Marphysa, The Thing broke into numerous fragments post-mortem, a detail Roesch confirmed with the divemaster at Anse Chastanet who reported recovering partial specimens on occasion.

The fact that multiple specimens have been recovered suggests a breeding colony, as opposed to one worm with gigantism. Marsden also informed Roesch that another giant polychaete, similar in description to The Thing, was rumored to exist in Barbados, suggesting The Thing is not limited to one island.

## Bermuda Blobs

In cryptozoological parlance, a "globster" is a mass of organic material that washes up on the shore. Whereas a carcass has sufficient skeletal features and/or body parts intact to identify the remains, a globster by definition is unidentifiable. The term was coined by Ivan T. Sanderson in 1962 to categorize the 1960 creature found in New Zealand and then applied retroactively to any historical report of an unknown mass that Sanderson was discussing. The most famous globster was the one that beached in 1896 on Anastasia Island off St. Augustine, Florida, the so-called St. Augustine Giant Octopus (Chapter 6).

In May 1988, a globster washed ashore in Mangrove Bay, Somerset, Bermuda. It was reported to Teddy Tucker, a legendary diver with more than 100 shipwreck discoveries credited to him, and whose work with scientists and naturalists has brought him international acclaim. He photographed and sampled the mass. Tucker's decision was prudent, as the globster later was washed back out to sea. For lack of a better name, it was just called the Blob.

Marine conservationist and author Richard Ellis interviewed Tucker after the Bermuda Blob discovery. Tucker described it as "2½-3 feet thick...very white and fibrous... with five' arms or legs,' rather like a distorted star" and estimated it weighed several thousand pounds. As is the norm with globsters, he reported that it had no bones or cartilage. Tucker also said it was "very dense and solid" and that trying to cut off samples was like "trying to cut a car tire." This difficulty in cutting the mass is also a trait of the globsters, the St. Augustine Giant Octopus reports mention this difficulty multiple times.

Tucker sent photographs to various marine experts, but no one could offer an identification. Local media ignored the discovery—considering the amazing lost treasures and exotic species Tucker had discovered, an unidentified mass was practically boring.

That changed in April 1995 when some of Tucker's samples would help resolve the identities of globsters. Sydney Pierce of the Department of Zoology at the University of Maryland performed electron microscopy and amino acid analyses on both the St. Augustine Giant Octopus and the Bermuda Blob. The result was that the globsters were pure collagen from known species that had been dead so long that predators and bacteria had rendered it down to collagen, the most resistant and least edible protein. The St. Augustine sample was from a shark; so was the Bermuda Blob.

A second, smaller globster (only weighing half a ton) was found in 1997, which the press again called the "Blob." It washed ashore near Surf Side Beach. This time, the Bermuda Natural History Museum was ready. A team of scientists descended upon the second Bermuda Blob and suggested it was a whale. There are two factors in this initial identification. First was Pierce's analysis two years earlier. Secondly, the texture of this blobster was more recognizable. As James Conyers of the Bermuda Aquarium reported to *The Royal Gazette*, "It smelled, looked and was the texture of decomposed whale blubber." He also noted that a decomposed whale had recently been discovered five miles up the coast in Devonshire Bay, which was missing "large chunks."

Large samples were taken and tested. The results were surprising. Although it was still assumed to be detritus from a cetacean, a new complication was added—the creature died after poisoning itself. The globster was found toward the end of a six-year epidemic of *Aspergillus sydowii* fungus attacking the soft coral known as the Sea Fan (*Gorgonia ventalina*). The whale would have been infected with aspergillosis after eating fish that ate sea fan predators. As one example, the whale eats a hogfish (*Lachnolaimus maximus*), which is a major predator of the flamingo tongue snail (*Cyphoma gibbosum*). The snail preys on the sea fans. As the whale consumes contaminated fish, the toxins build up until the whale develops fatal pulmonary aspergillosis. Bermudian marine biologists knew what killed the animal, even if they weren't still sure what the animal was.

In July of 2003, another globster was discovered in Chile. Pierce immediately received a sample, taken from the globster in situ. Pierce had gotten so much pushback from livid cryptozoologists that he decided to resolve the matter. The sample, as well as samples from St. Augustine, both Bermuda Blobs, and others found across the globe, were tested. The results were that all the samples were collagen. And the Chilean sample yield sperm whale DNA. The globsters were decomposed remains of known aquatic creatures. (Pierce's analyses are discussed in Chapter 6).

Globsters continue to wash up on beaches. Local newspapers use them as "evidence" of sea monsters; these stories, in turn, are disseminated and perpetuated on the internet. Apparently, globsters, being made of collagen, are resilient.

# CHAPTER 5
## Krakens of the Caribbean

In 1802, Pierre Dénys de Montfort, a French malacologist, published a supplemental volume to Comte de Buffon's *Histoire Naturelle Générale et Particulière*. This volume was dedicated to mollusks in general, but de Montfort includes two giant cephalopods, the colossal octopus, and the kraken octopus. The decision to include these two species was based on de Montfort's interviews with fishermen and whalers. De Montfort considered both octopuses to be aggressive sea monsters, probably responsible for many of the ship disappearances over the years. This is also where the confusion began over what precisely a kraken is.

The earliest published use of the name "kraken" itself appears to have been by the Bishop of Bergen, Erik Ludvigsen Pontoppidan, in his *Det Förste Forsög paa Norges naturlige Historie* (1752) with the 1755 English translation *The Natural History of Norway*. Pontoppidan also collected his information from the fishermen and sailors, indicating the term was already in common usage. The etymology of the word remains debated, but the most prevalent theory is that it is derived from the adjective *krake*, referring in a Norwegian dialect to an unhealthy, unnatural, or "twisted" thing. Pontoppidan actually couldn't describe the creature—the beast's immense size and underwater habitat made even the most observant fisherman unable to get a clear view. The bishop opined that the kraken was dangerous but not aggressive. Deaths and injuries were accidental, caused by fishermen unable to get out of the kraken's way when it surfaced or in the whirlpools created when it submerged.

Pontoppidan suggested the great beast ate so much fish that when it defecated, the water was clouded by excrement so high in fish content that it would act as a lure to attract more fish to meet their fate. From the descriptions he gathered from the docks, Pontoppidan concluded the kraken was a giant starfish.

De Montfort included an illustration of his colossal octopus pulling a boat under the water, probably based on mistranslating Pontoppidan's reference to ships being caught in the undertow. De Montfort's designating the kraken as a cephalopod might also be a translation error, with Pontoppidan's reference to scatological murkiness being interpreted as the release of cephalopod ink.

By the 1850s, the kraken was synonymous with a giant squid, thanks in no small part to a series of papers written by Danish zoology professor Japetus Stéenstrup, who assigned it the name *Architeuthis dux* (ruling squid). Stéenstrup based his work on samples and specimens; it wasn't until 1861 that the existence of the giant squid was confirmed when a portion of one was secured by the French corvette *Alecton* in 1861, followed by a series of strandings in Newfoundland and New Zealand.

Thanks to Walt Disney and Ray Harryhausen, the kraken has become a pop favorite, although the movies still use the octopus and squid interchangeably.

## The Giant Scuttle

Forrest G. Wood, Jr., better remembered for his part in reviving interest in the St. Augustine giant octopus, had approached that particular story with a touch of confirmation bias. He had found the newspaper clipping that piqued his interest just months after a March 1956 trip to West End on Grand Bahama Island, scouting locations for the acquisition of specimens as Director of Exhibits at Marineland in St. Augustine.

In the third part of the March 1971 *Natural History* article he co-authored with Dr. Joseph Gennaro, Wood recalled that Duke, his fishing guide on that trip, had a comprehensive knowledge of the area and the aquatic denizens off the coast. As they were finishing up one evening, Wood recalled a Bahamian folktale he heard in his first job, just before Marineland, at the Lerner Marine Laboratory on Bimini, the Bahamian island 60 miles south of their location. On a whim, Wood asked Duke if he had heard the tale of a giant "scuttle fish," scuttle being the local name for an octopus.

To Wood's surprise, Duke not only knew of the giant scuttle, he rattled off a list of witnesses, including date and location, noting that the last report took place a decade previously. Duke claimed these scuttles had tentacles more than 75 feet in length but only came into shallow water when sick or dying. Wood was skeptical, even with specific names, locations, and dates provided. He was less skeptical a few days later when he met with British Colonial Commissioner Stanley R. Darville. Darville was of old Bahamian stock—his French-Scottish ancestors had arrived in the 17th century, and he was essentially the British government on the island, handling everything from immigration to accident investigations to vital records.

When Wood asked Darville about scuttles, he told a tale from when Darville was about 12-years-old (circa 1926). He had been handline fishing off Andros with his father and a family friend in 600 feet of water. They were fishing for silk snappers. However, Darville's father hooked something substantial. Assuming they had snagged the bottom, they attempted to pull up the line. It came up, but slowly, as if carrying a great weight. When the end of the line came in sight, they could see a "very large octopus" clinging to it. The cephalopod detached itself from the line and attached itself instead to the boat's hull. The octopus then lost interest, to the relief of the panicked passengers. Wood asked Darville specifically about the size of the octopus in the encounter. Darville would not commit to an estimate after three decades but made it clear he was not talking about one of the common octopuses that inhabited the shallow waters.

As an aside, it is unfortunate to note that Wood did not pursue the 10-year lapse in scuttle sightings. The end of sightings appears to correspond with a 1947 massive red tide algae bloom in the Gulf of Mexico that lasted nearly a year. Starting near Florida's panhandle, the infected area extended far enough south along the coast for the Gulf Stream to have carried sufficient diluted toxins along the Gulf Stream to the Bahamas to disrupt breeding cycles or encourage a cephalopod migration.

Wood contacted Gilbert Voss at the Marine Laboratory of the University of Miami. Voss was already considered an expert on the zoogeography of cephalopods, particularly those of the Caribbean Sea. Voss replied that he was well aware of the Bahamian scuttles but thought there might have been some exaggeration in the encounters as retold to Wood.

Wood mused that the term "scuttle fish" was a portmanteau noun made by combining cuttlefish and the cephalopod's scuttling motion. Others have noted that scuttle is the verb used to describe the act of sinking a boat. The term is old English, appearing as "scottell a fyssche" in John Palsgrave's French grammar textbook in 1530 London. The term "scuttle" to describe an octopus appears in passing as early as 1812 in a poem by missionary Joshua Marsden, indicating it was in common usage in Bermuda, but qualitative usage doesn't appear in zoological texts until Pickford's *Le Poulpe Américain: A Study of the Littoral Octopoda of the Western Atlantic* in 1945 where she states that the common octopus, (*Octopus vulgaris*) is called the "Rock Scuttle" in Bermuda while the grass octopus (*Octopus macropus*) is called the "Grass Scuttle." The first colonists arrived in Bermuda in 1612, 82 years after Palgrave's book, so it appears the term arrived in the Caribbean via Bermuda and spread to the

Bahamas in 1647.

How large a Giant Scuttle is depends on whom you ask. Bahamian hardware store chain owner John S. George of Nassau wrote to Professor Burt Wilder at Cornell in 1872 telling of a 10-foot long octopus that had washed up on the beach. George notes it was the first one he had seen in 27 years as a resident in the Bahamas, but there was a local tradition of octopuses of "immense size." In 2011, the *Freeport News* chronicled the mostly depredated remains of an octopus that had washed ashore on William's Town Beach; it was estimated to have between 20-30 feet long, comparing it to a July 2010 fishing boat that found a large octopus floating off the Bell Channel. That octopus weighed in at 135 pounds and measured 12 feet. These sizes are not typical of octopean species in the Caribbean. The giant Pacific octopus (*Enteroctopus dofleini*), the largest known octopus species, averages around 33 pounds with an arm span of upwards of 14 feet.

In a 1983 interview for the International Society of Cryptozoology, Forrest Wood stated his belief there was a similarly large but unknown octopus in the deep waters off the Bahamas. The sightings suggested that much like the *Enteroctopus dofleini* in the Pacific, this Caribbean cryptid would be found on the sides or bases of steep slopes, such as the "Tongue of the Ocean" on the east side of Andros Island. His concern with the theory was the availability of food sources. Most octopuses feed on crustaceans such as crabs, and Wood was unaware of any large crustaceans in that area.

Wood's concern may have been allayed by a 1985 issue of the *International Society of Cryptozoology Newsletter*, which reported on the troubles of Sean Ingham, owner of Pathfinder Fisheries in South Hampton, Bermuda. Ingham had made his fortune before Bermuda began to tighten regulations on the number of traps allowed. Ingham was experimenting with specialized traps, capable of reaching depths down to 6,000 to 12,000 feet (1,000-2,000 fathoms). Ingham's traps were getting better results than expected. The traps yielded large shrimp and two different species of Geryon crabs. Golden crabs (*Chaceon fenneri*) was already being fished commercially in the Gulf of Mexico but was unknown in Bermuda. The other was a new species, *Chaceon inghami*, named after the discoverer. And these crabs were large—in the five to six-pound range and at least one with a carapace about 20 inches across.

Starting in July 1984, Ingham had been experimenting with eight-foot by eight-foot, four-foot-high trap. The trap was a common style of traps in the Caribbean, an Antilles arrowhead, named so because of the chevron entry point. Ingham's traps were customized for deep water, constructed of two-inch

hexagonal galvanized mesh wire across frames of steel reinforcing rods ranging from ¼ to ⅜ inch in diameter, cross-braced with two-inch-thick wooden
staves. He moved this trap up and down the slopes of the areas between the
banks and Bermuda's edge. At 420 fathoms in an area where the currents were
particularly strong, he pulled up a trap that was virtually packed with crabs.
The combined weight of 5000 lbs. in crabs and the trap, coupled with the current, was too much for the winch. Struggling to get the trap aboard, it broke
apart against the side of the boat. So he then started using smaller traps. He
noticed some traps were coming up damaged, which he assumed was caused
by a laden trap being dragged along the bottom as it was being raised.

On September 3, a second trap was lost. A smaller, six-foot by six-foot,
three-foot-high trap at 500 fathoms was being raised that seemed significantly
heavier than it should have been. At 300 fathoms, the capstan winch was
pulled backward, popping the rope off the capstan. Unencumbered, the line
flew off the deck at 30 mph. Ingham was able to guide the rope back around
the capstan while his partner hosed down the capstan head to minimize friction. The winch was engaged at a higher torque and the trap began rising
again. At 250 fathoms, a series of sudden jerks shook the line and it snapped.
It would take a weight of over 600 pounds to break the line, which was poly-
ethylene rope.

On September 16, *Trilogy* was trying to bring up small, three-foot by
three-foot, one-foot-six high rap from 480 fathoms but the winch could not
lift it off the bottom. The winch was beginning to strain, trying to hoist against
the equivalent of 4,500 lbs. of weight. Ingham checked his chromoscope—a
downward-pointing sonar customarily used as a depth recorder, and saw a
"pyramid shape, approximately 50 feet high" on the trap. Whatever it was,
it was the same size as the 50-foot *Trilogy*. After 20 minutes of trying, the
unknown creature began slowly towing the boat to the south. After a ⅓ mile,
it was apparent they were heading back toward the inshore shelf, where the
topography is riddled with depressions and deep holes. Ingham held the line
and was able to feel a pattern of vibrations—a pattern that reminded him of
something walking. The rope went slack and the trap was quickly winched
aboard, with only slight damage to one side.

The loss of the two traps (and possibly the original eight-foot one) made
Ingham suspect some sea creature, hundreds of fathoms deep, was attracted to
the crabs, intelligent enough to recognize the traps as the source, and strong
enough to break the line. A giant octopus seemed the most likely candidate.

While there are unquestionably traditions and suspicions of giant octopuses interacting with vessels and fishing operations across the Caribbean, only on Andros do these aggressive scuttles have their own distinctive folklore and name—the lusca.

## Lusca: Beast of the Blue Hole

The lusca is half-shark/half-octopus, half-eel/half-octopus, or half-dragon/ half-octopus, depending on who you ask. The key term is half-octopus, for the lusca is said to pull fishermen and their boats underwater, either by its giant tentacles or by inhaling so powerfully to create a whirlpool, exhaling the inedible remnants as flotsam. They are said to inhabit blue holes on Andros Island. The lusca is not a myth retold across the Bahamas. The story of the lusca only began to spread off Andros in the 1950s when the Bahamas began to develop into an American tourist alternative to Cuba and anglers began hiring local Androsians as guides.

Blue holes are not exclusive to the Bahamas, although Andros has more proportionally than anywhere else. The term itself is of local origin, derived from the deep blue color indicative of the cave entrances. Such caves holes are typically found in on or near islands composed of a carbonate bedrock, such as limestone or coral reef. It is essentially a sinkhole dating back to the Ice Age that extends below sea level into submerged cave passages. The inland blue holes are deep enough that they are among the oldest stable environments in the world, with their own ecosystems and specialized species.

Tides significantly influence blue holes. At high tide, the ocean is higher than inside the cave, so as the water level inside the cave attempts to balance, a vortex is created as it draws water through the cave mouth opening. The larger blue holes create currents strong enough to keep a diver from escaping a cave and drawing down anything floating on the surface, such as an overloaded small fishing boat. Similarly, at low tide, water, bubbles, and captured debris are disgorged in a visible bulge on the ocean surface. These marine blue holes are sometimes referred to as "blowing" or "boiling," depending on whether the lusca is inhaling or exhaling.

Envisioning a scenario where a fisherman in a small boat is suddenly caught in an eddy from such a marine cave, it would not be a great leap to assume the whirlpool is from an attacking lusca. With the "giant scuttle" legends and the number of verified remains of Giant Squids being found floating on the surface around the islands, it would be easy to assume the lusca is a tentacled monster. Between 1952 and 2011, the Smithsonian's Clyde Roper inventoried 15 confirmed sighting or carcass retrievals from the Straits of Florida and Bahamas; the addition of tentacles to the lusca could just be an embellishment to what was already a very good story. This is not to suggest giant squid are or aren't competing with the scuttles for a meal among the fishing community of the Bahamas, but unrelated drownings and missing vessels without eyewitnesses only add to the lusca legend.

On the southeast side of Andros are four major marine blue holes in the area of Grassy Cay, including the "Lusca's Breath," the third deepest cave known in the Bahamas. The cave also has some of the most powerful currents, creating a significant whirlpool that develops at the mouth during high tide. Considering the entrance shaft descends 150 feet before extending out in multiple directions, anyone or thing caught in the vortex will not survive.

The original Native Bahamians, the Lucayan tribe, had to be aware of the blue holes, but whether they associated them with a creature will never be known. From a peak population of about 40,000 at Columbus' arrival, the Lucayans were decimated and exploited by the Spanish—enslaved, sent to other islands, and ravaged by disease. By Ponce de Leon's visit to the islands in 1513, the Bahamian islands were uninhabited.

In 1647, the Eleutherian Adventurers, Puritans who left Bermuda to settle on the island of Eleuthera, arrived. Britain officially took over the colony in the eighteenth century. At the end of the American Revolution, the Bahamas saw an influx of Loyalists from the fledgling United States—and they brought their slaves. Today close to 85% of the Bahamas are descended from slaves with ancestries across Africa. The Bahamian folklore, like much of the Caribbean, reflects a West African origin influenced by other cultures, such as English, French, and Spanish, depending on which island you're discussing.

The term lusca remains specific to Andros and a mystery as to its origin. Bahamian dialect is heavily influenced by Gullah, similar to Sierra Leone Krio, a dialect used by Carolina and Georgia slaves brought by Loyalists fleeing the American Revolution (Britain did not outlaw slave trade until 1807). Most settled on Andros, further complicating the potential source of the term. The Carolinas, particularly around the Cape Fear River valley in southeastern North Carolina, had a large community of monoglot Highland Scots—an estimated 20,000 Gaelic speaking Highlanders who had emigrated between 1763 and 1773. By 1790, about a quarter of these Highlanders had an average of five slaves, who early travel writers noted, were also fluent in Gaelic. So, it is interesting to note that with Scottish and Irish among the early settlers in the Bahamas, an influx of American Loyalist slaveholders, and later escaped slaves, there was ample opportunity for some Gaelic to be introduced into the vernacular.

In Gaelic, *lusca* means a cave, or underground dwelling, which might suggest the blue hole itself was originally named lusca, not the inhabitant.

### The Bimini Beast

In late October of 1963, something large was spotted in deep water off the Bahamas by a fishing charter. It was not the first sighting, but this time, the Miami papers noticed. Burton Clark, the general manager of Miami's Seaquarium, told the *Miami Herald* that marine experts thought it could be either a giant squid or a prehistoric shark.

Burton Clark, it is important to note, had been employed by Wometco Entertainment Group as a theater manager in Indiana for 25 years when the company, a major theater chain in South Florida, brought Clark to Miami in 1960 to turn around Wometco's struggling marine tourist attraction. Clark's job was to make the Seaquarium into a major tourist attraction, which he did successfully. And one of his marketing ploys was to send the company's vessel out to investigate sea monsters, so his marine experts may have been part of the attempt to generate interest. And in 1963, *20,000 Leagues Under the Sea* had returned to Miami as a drive-in feature, so Clark was going use the giant squid association, regardless of what the cryptid actually turned out to be.

Don McCarthy, the director of fishing information for the Bahamas Development Board, had a different story. The most recent witness, a seasoned captain, said that the mystery beast was no mystery—it was a whale shark (*Rhincodon typus*), the gentle giant filter feeder. McCarthy noted they had caught a whale shark four-to-five years previously. Gilbert Voss at the University Marine Laboratory, a cephalopod expert, also thought it was a whale shark.

Be it squid or whale shark, Bahamian charter boat captains had been reporting the creature for several years, something "big and black with brownish-yellow spots" that had once nearly capsized a 40-foot cruiser. The reports appeared to center around one specific section of the sea—the newspapers used Bimini as a landmark, placing the sightings 70 miles south of Bimini, but that position would be barely 50 miles off the westernmost tip of less tourism-driven Andros Island.

The Miami Seaquarium had been approached by the sports editor of Time/Life publishing to help investigate the monster fish and agreed to send the yacht *Seaquarium*, their collection boat, out to investigate that summer. Twelve days later, the expedition ran out of time and money—the Seaquarium had overspent preparing a dolphin show for the upcoming World's Fair in

New York.

In 1975 cryptozoologist Gary Mangiacopra corresponded with Clark, who shared details not included in the newspapers. Clark noted the expedition's lack of success meant no logs or readouts were kept, but that sonar contact had been made with something large on the ship's fathometer. The ocean was 1,000 feet deep, and whatever appeared on the equipment was at 600 feet. The other detail of interest is that part of the Time/Life team aboard the *Seaquarium* was Dr. Harold Edgerton, the inventor of the strobe light, who was testing a new submersible camera with a strobe-flash triggered by a bait line. So the Bimini Beast may have been a bust, but Edgerton would continue to improve the technology. Although it would never photograph a kraken, it captured an image just as impressive. In 1972, Edgerton's strobe camera took the famous "flipper photo" at Loch Ness.

# CHAPTER 6
## The St. Augustine Giant Octopus

On Sunday, November 29, 1896, 19-year-old Herbert Colee and 18-year-old Dunham Coxetter were bicycling on Anastasia Island, the barrier island separated from the St. Augustine mainland by the Matanzas River. The weather had been terrible since September. A hurricane had crossed the Bahamas early in September, and a second passed off the Florida coast on September 28. There had been no damage, but the endless clouds and rain put a damper on most outdoor plans. Adding insult to injury, a late-season hurricane then came ashore near the Gulf Coast village of Cedar Key on October 9 and continued northeast across Florida, passing north of Jacksonville before reaching the Atlantic and heading up the coast. This time, Saint Augustine saw damage from gale-force winds with gusts up to 50 miles per hour.

So with storm cleanup finished, Colee and Coxetter had spent a lazy Sunday riding their bicycles on the hard-packed sand of Anastasia Island. It would be one of their last opportunities to take a tranquil ride along the beach and explore the ruins of the old Spanish fort at the Matanzas Inlet, as snowbird season had begun and the northern crowds were descending on St. Augustine. After investigating what the storm might have uncovered at the fort, they spent the rest of the afternoon in the surf along South Beach, a generic name for most of the beach south of the St. Augustine Inlet, now called St. Augustine Beach. Then, on a stretch of deserted shore known today as Butler Beach, they spotted something half-buried in the sand, shapeless but obviously organic. The exposed part was big, and indications were that there was more buried beneath the sand.

Based on the sheer size, their first thought was that it was the remains of a whale, washed ashore by the recent hurricane. A whale carcass had washed ashore at the Matanzas Inlet just two years earlier, creating much excitement among tourists and locals alike. So they knew what to do. They headed back to Saint Augustine to the King Street home of Dr. DeWitt Webb.

DeWitt Webb was a respected physician, a skilled politician, and an amateur naturalist.

The son of Hudson River Valley farmers with roots back to the American Revolution, his health had failed from the strain of running a part-time medical practice, operating a wholesale pharmacy, and being a civic leader in Poughkeepsie, New York. After a term in the New York State Assembly, he and his wife, Adele, moved to St. Augustine; in 1880, the climate was being touted as being a panacea for whatever ailments afflicted you.

It apparently worked. By 1896, Dr. Webb had recovered fully, and over the next 16 years became a community leader. With the arrival of the Flagler railroad and hotels, the year-round population leaped from 2,300 to 5,000. Once snowbird season started, the population increased annually, fluctuating week to week. Webb found his leisurely schedule of civic duties, medicine, and antiquarian pursuits hampered by the massive spike in the tourist population. There were only six doctors in town, plus one who lived on hotel property to serve as 24-hour medical access to Flagler's three hotels, exclusively handling the ailments for the cream of society. If the ailment was more serious than a sprain or a sunburn, the hotel doctor usually referred the case to Webb or one of the four other doctors who were also licensed as surgeons.

Most of the year, Webb's medical practice left him time to run the Historical and Scientific Society he had founded, volunteer at the segregated "Black School," tend to Apache prisoners of war housed at Fort Marion, dig into burial mounds for Indian specimens for the Smithsonian, and tend to his business investments, both in Poughkeepsie and at his Florida citrus groves. His business investments were a necessity with six doctors dealing with such a small residential population of varying degrees of income.

When Coxetter and Colee came into his home/office claiming a whale carcass had washed up on South Beach, Webb took them seriously. He knew Colee well—his father was the town tax collector and his other relatives were on the City Council, Police Commission, and School Board. But it was already late Sunday afternoon, and Webb was active in the church. He couldn't drop everything and head over to the beach to examine a rotting whale carcass. The previous whale carcass had been in too poor of shape to remove the skeleton for the Historical and Scientific Society, and Webb wanted to add the skull to the collection, but he had done enough specimen collecting to know a whale carcass wasn't going anywhere overnight.

By Monday morning, November 30, 1896, word began to get out about the whale on South Beach. By the time Webb arrived with fellow members of his Saint Augustine Institute of Science and Historical Society late that afternoon, there was already a steady stream of curiosity seekers, including a newspaper reporter. Webb, whose zoology background was limited to collecting fish specimens for the Smithsonian, made a cursory examination in deepening shadows and decided it was not a whale—it is an enormous octopus missing its tentacles. The next day's *Florida Times-Union* announced the discovery as "Big Octopus on the Beach":

> St. Augustine, Fla. November 30. A great fish, supposed to be a whale, 22 feet long, 8 feet wide and about 6 feet high, was found partly buried in the sands of Anastasia Beach yesterday evening by Herbert Colee and Dunham Coxetter who were making a cycle run to Matanzas Inlet. The immense carcass lies about four miles south of Dr. Grant's hotel and has probably been beached several days, as it is well embedded in the sands: it is where a whale was caught in the Matanzas River two years ago. President Webb of the scientific society examined the monster this evening and says it is an octopus. The tentacles are gone, either by the action of the sands or through sharks eating them off.

A point of interest is the geographic landmark used to identify where the carcass beached: "Dr. Grant's hotel," not the terminus of the St. Augustine and South Beach Railway, which happened to be next to Grant's hotel, the Casa Marina. This suggests the first person to contact the press was Dr. Grant himself. Grant was a former physician who had decided there was no money in medicine and became a hotelier. Any question as to whether Grant was active in generating interest in the sea monster was settled with the next news-paper article that appeared. The *New York Herald* of December 2 includes a telegraphed report of the sea serpent on the beach with no mention of Webb. It does announce Dr. Grant had "secured the monster and will try to preserve the skin and skeleton," and that the identification of the species might be pos-sible when the buried portions were unearthed. If Dr. Grant did not send the report to New York himself, there was another possibility—Dunham Coxetter had just started his career as a telegrapher.

In 1896, Anastasia Island and the beaches were much more convenient to visit than they had ever been before, courtesy of a new toll bridge that crossed the river to the St. Augustine and South Beach Railway depot. A horse-drawn passenger car carried tourists on rails along the western edge of Anastasia past the various attractions—the ruins of an old Spanish lighthouse, the current lighthouse, the coquina quarries—and ended at South Beach where a small collection of winter cottages, a store, and Grant's hotel were located. Grant also maintained the South Beach Pavilion next to his hotel. Grant had been waiting for the railway to complete the bridge across the Matanzas, replacing a ferry that just went back and forth hourly. Flagler's hotels in St. Augustine had taken most of the potential customers from the Grant and his Casa Marina, as did the other hotels that sprang up on the mainland. So, the remains of a sea serpent washing ashore four miles south of his location must have seemed a godsend. Curiosity seekers would leave the railway and walk the final four miles (or rent a carriage from Grant for a small fee) down the beach to the sea monster. Folks were returning to the railway hot, tired, and hungry. Already the little general store and the evening fish fry at the pavilion were gaining in popularity.

The carcass had probably been on the shore for several days by the time Colee and Coxetter stumbled across it. It was assumed to be a recent arrival because of the condition of the remains, but as the carcass remained on the beach over time, the minimal decomposition exhibited could very well indicate it washed ashore any time after the October 9 storm. The estimated weight of five-to-seven tons had caused it to sink into the sand. The portion above the sand measured 23 feet in length, 4 feet high, and 18 feet across the widest point, with an unhealthy pinkish tint that turned into a silvery gray color.

Webb would not return to South Beach until December 5. His initial interest had been in securing the whale's skeleton for the museum. A big octopus was interesting but certainly not as high a priority. In between visits, he had discovered that this octopus was either a record-setting size or a new giant species. By the time he returned, he had lost control of the situation. Dr. Grant was trying to move the "sea monster" to the South Beach pavilion to exhibit (for a small fee), and curiosity seekers were attempting to cut souvenirs off the carcass—unsuccessfully, however, as reports stated that the beast's hide was impervious to pocket knives. This time, Webb took more extensive notes and then came back two days later, accompanied by local photographers to document "his" octopus. While the photographs were being developed, Webb began a series of letters to various scientific journals and zoologists, notifying them of the discovery. With minimal variation, the letters read the same:

> You may be interested to know of the body of an immense Octopus thrown ashore some miles south of this city. Nothing but the stump of the tentacles remain, as it had evidently been dead for some time before being washed ashore. As it is, however,

the body measures 18 feet in length by 10 feet in breadth. Its immense size and condition will prevent all attempts at preservation. I thought its size might interest you, as I do not know of the record of one so large.

The earliest surviving letter, dated December 8, was sent to Joel Asaph Allen of the American Museum of Natural History in New York. Dr. Allen was undoubtedly a bit confused by Webb's letter—he was best known for his work in ornithology and was the curator of birds and mammals at the museum. He passed the note along to the curator of the museum's geology and fossil collection, Robert Parr Whitfield, an expert in invertebrate fossils, including mollusks. Whitfield didn't know what to do with the query either and forwarded it to Addison Emery Verrill, Professor of Zoology at Yale University in New Haven, Connecticut.

Verrill should probably have been the first person Webb contacted, but no note reached him until Whitfield sent him a copy. A prolific writer who classified thousands of new species, Verrill was particularly interested in giant cephalopods. In 1871, he had cemented an international reputation by legitimizing scattered reports of giant squids and establishing a classification system for giant squid species.

Just as important, Verrill was decidedly open-minded to the possibility of a giant ocean species. In an article in an 1875 issue of the *American Naturalist,* he had noted that the gladius of the giant squid seemed to resemble that of the *Teudopsis* genus, a squid found only in fossils of the lower Jurassic period, and "contemporaneous with the huge marine saurians, *Ichthyosaurus, Plesiosaurus,* etc., the 'sea-serpents' of those ancient seas. May there not also be huge marine saurians still living in the North Atlantic, in company with the giant squids, but not yet known to naturalists?"

As Webb proceeded carefully, arranging for additional photographic documentation, George Grant's media blitz had gone national, now giving Webb and Grant equal credit as protectors of the sea monster. One article made its way across the country's newspapers with Grant's fanciful description of body parts that didn't exist (such as a "jagged tail"). One such reprint appeared in the statewide *Pennsylvania Grit,* legitimizing the description across the entire state, and generating additional reprints of the story nationally.

Verrill, having received the description from Webb, jumped the gun and published Webb's report in the January 1897 *American Journal of Science.* He suggested the carcass was actually a squid but even bigger than the giants he had examined. After receiving photographs from Webb, Verrill reversed his opinion, deciding that it was indeed a gigantic octopus.

Sometime between January 9 and 15, a new problem arose for Webb. The approaching full moon was pushing the tides higher, and the ocean was encroaching on the remains. Combined with a series of typically Floridian sudden storms, the inevitable happened—the carcass was washed back out to sea. Fortunately for Webb, it washed ashore again, but it was now another two miles further down the shore, at Crescent Beach. The new location all but necessitated renting transportation from Dr. Grant's hotel. No doubt it chagrined Webb to see Grant profiting from the discovery when Webb was spending his own money trying to move the octopus above the high watermark. The first attempt, with a dozen men using block and tackle, was unsuccessful, and the workers were unable to even flip the remains over.

Webb wrote letters to William Healey Dall, Curator of Mollusks at the National Museum at the Smithsonian in Washington, D.C., for advice and included less than subtle hints that funding would be greatly appreciated. On January 16, on a second attempt, Webb was able to move the carcass 40 feet from the shoreline, well above any high-water marks. As he later wrote to Verrill and Dall:

> Yesterday I took four horses, six men, three sets of tackle, a lot of heavy planking and a rigger to superintend the work and succeeded in rolling the invertebrate out of the pit and placing it about 40 feet higher up on the beach where it now rests on the flooring of heavy plank. After getting it out we found it on being straightened out to measure twenty-one foot instead of eighteen as I first reported to you. A good part of the mantle or head remains attached near to the more slender part of the body. This was spread out as much as possible. The body was then opened for the entire length of 21 feet as you will see by the new photographs yet to be sent. The slender part of the body was entirely empty of internal organs. And the organs of the remainder were not large and did not look as if the animal had been so long dead as it appeared to have been when first washed ashore some six weeks since. The muscular coat which seems to be about all there is of the invertebrate is from two and three to six inches in thickness. The fibers of the external coat are longitudinal and the inner transverse. There was no caudle fin or any appearance as if there

had been any. There was no beak or head or eyes remaining. There was no pen to be found nor any evidence of any bony structure whatever…

Webb dropped a quick follow-up note the next day advising Dall of the newest photographs and complaining he was forced to make yet another trip to the beach. Word had gotten back to him that Dr. Grant had decided to move the sea monster back to the South Beach Pavilion for display, possibly by cutting it into more manageable pieces. Webb suddenly found time to visit the carcass. He erected stakes and roped it off, posting legal ownership on behalf of the Saint Augustine Institute of Science and Historical Society. Grant was prevented from moving the remains without a court battle, but Webb ended up looking rather petulant.

It would take another three weeks for Webb to return, this time to secure samples for Verrill and Dall. In a February 5 letter to Dall, Webb notes he has cut several samples that he would soak in formalin for a few days before shipping them. At the end of the letter, Webb announced he had a plan to move the carcass six miles up the coast to the end of the railway, where it would be exhibited, with the small viewing fee going to a man who will maintain the remains. In other words, Webb would allow George Grant to implement the plan he had been proposing since November. The teams of horse and tackle financed by Webb personally and had been so expensive that the cost of preserving the entire mass was prohibitive to Webb's purse. The compromise was to modify Grant's plan so that a portion of the profits would go toward underwriting the preservation of the remains.

Meanwhile, Verrill had submitted an update to the *American Journal of Science* that ran in the February issue that recapped Webb's work and included the revised dimensions as measured during the move. It also included the estimated tentacle length, though Webb carefully neglected to mention that someone else had measured the tentacles and that he hadn't even seen the now-missing appendages.

By the end of February, both Verrill and Dall had received their samples. That would prove to be the death knell for Webb's "giant octopus." Verrill broke the news to Webb in unarchived correspondence: Verrill wasn't sure what the sample was from, but he was adamant it wasn't from a cephalopod. Dall helpfully suggested it was probably from a whale. The news was not well received. Webb wrote one last letter to Dall in the middle of March, asking about purchasing formalin in bulk, still planning to preserve the specimen for the historical society. He also took the time to lament the revised identification, simply unable to see how "a great bag" could be part of a cetacean.

Verrill had already begun dismantling the Gigantic Octopus identification and salvaging his own reputation in the process. In the March 19 issue of *Science*, Verrill emphasized that his mind was changed by the samples Webb had sent. He noted the samples were composed of strong elastic connective tissue fibers like those of cetaceans, not the highly contractile muscular tissue indicative of a cephalopod. Immediately following Verrill's article is a brief addendum by Frederic A. Lucas, a Smithsonian naturalist who had just spent months at a Newfoundland whaling station researching whale anatomy. He was less delicate:

> The substance looks like blubber, and smells like blubber and it is blubber, nothing more nor less ... The imaginative eye of the average untrained observer can see much more than is visible to anyone else.

Even the *Tatler*, the society's local newspaper, reported Verrill said it was no octopus, essentially ending the stream of tourists to see the carcass. And with the drop in tourism came the decline in revenue. By the end of spring, the carcass was quietly buried in the sand in situ.

In April, Verrill continued to back away from his previous claims. His article in the *Annals and Magazine of Natural History* was bluntly titled "The supposed Great Octopus of Florida; certainly not a Cephalopod." A final article by Verrill in *American Naturalist* starts out almost apologetic to Webb and goes into the specifics of why the samples could not be from a cephalopod. Verrill suggested it was part of a sperm whale's misshapen head after the skin sloughed off. He added a footnote mentioning that the spermaceti case of the sperm whale was also a strong candidate. The Great St. Augustine Sea Monster faded into obscurity—for a while.

The octopus/whale matter did not impact Dr. Webb's civic standing. He immediately returned to devoting his spare time to such diverse interests as collecting Indian artifacts, becoming president of the Florida Medical Association, and doing a stint in the state legislature. In 1911, Webb was elected mayor of St. Augustine. In 1914, the headquarters of the St. Augustine Historical Society and Institute of Science burned down, destroying any records or samples of the carcass not already distributed. He died in 1917 after three years of rebuilding the historical society. The Webb Building of the St. Augustine Historical Society's Oldest House complex was built in 1923 and named in honor of his 34 years as the group's president.

Dr. George Grant continued to run the Casa Marina. He was named a director of the St. Augustine and South Beach Railway, which had been struggling financially after upgrading to a steam locomotive. The railway slipped into receivership and was purchased in 1904 by the St. Johns Electric Company, who electrified the route and rolled it into their city trolley system. Still involved in the railway, Grant gave up the hotel and moved back to St. Augustine proper. In 1911 he moved down the coast to Ponce Inlet to run another hotel, where he died in 1913.

Dr. Webb's giant octopus became a footnote in zoology but continued to generate passing mentions, albeit with diminishing frequency. A. E. Verrill's son, A. Hyatt Verrill, referenced the beaching in his 1916 book, *The Ocean, and its Mysteries.* The son's opinion, only 20 years after his father's final pronouncement, opines the opposite—that the beast was an unidentified species. A. Hyatt was a witness and a participant in his father's work—his drawings from the overexposed Webb photographs accompanied his father's work.

At the Smithsonian, naturalist Fred Lucas made one last comment on the matter in 1928, using it as an example of how not to approach research, mocking A. E. Verrill by recalling a dreadfully British comment about the risks of identifying Florida specimens while sitting in one's study in Connecticut. Considering that particular bon mot had run in the British journal *Natural Science* in May of 1897, Lucas had been waiting for 30 years to use it. Lucas also identifies the St. Augustine monster as detritus from a whaling vessel, the "wave-worn case of a sperm whale, from which the spermaceti had been taken before it was cast adrift."

Minor reference aside, the Giant Octopus was then mostly forgotten until the 1950s. A. Hyatt Verrill returned to the topic in another book, *The Strange Story of Our Earth,* one of the last of his 105 books before his death in 1954. In this book, the younger Verrill still believed the remains were of a completely unknown creature but added other recollections such as the fact his father was the one who sent tanks of alcohol to Webb to ship the samples. This suggests that Verrill's memory was not reliable after 55 years—formalin (formaldehyde), not alcohol, was what Webb had been seeking. Webb's difficulty in obtaining enough to preserve the remains was because formalin was not introduced to the US retail markets until 1897, meaning there was no domestic supply for Webb to acquire. The Smithsonian had been importing formalin from Germany as early as 1881 as a preservative for brain tissue, but in 1897, formalin was still expensive and scarce. Neither the Smithsonian nor Verrill would have any supply readily at hand, particularly in the quantifies Webb would need. Webb had no choice but to abandon plans to save the entire mass as importing enough formalin from the German manufacturer to immerse a seven-plus ton mass would have been prohibitively expensive. A few more months would have made all the difference; later that year formalin was introduced to the US markets and became a wildly popular fad, used for treating grain fungus, as a meat preservative, as an inhaler for sore throats, and even briefly as an injection to cure blood poisoning.

It does raise the question of what the samples sent north were actually shipped in—alcohol, formalin, or some combination concocted by Webb. Dr. Joseph Gennaro, who would later analyze Webb's samples, found the pieces of the creature soaking in what he thought was a pungent combination of formalin, seawater, and alcohol.

Webb's giant octopus may have been buried on the beach, but it would not die quietly. In 1951, Forrest G. Wood, Jr. was hired at Marineland as a curator. Located on Anastasia Island, Marineland was initially opened as Marine Studios, an underwater film studio. It had always been popular as a tourist attraction; when trained dolphins were added in the early 1950s, Marineland became one of Florida's major attractions with filming delegated to a secondary role. With his background as a biologist at the Lerner Marine Laboratory on Bimini, Wood quickly rose through the ranks and became Director of Exhibits at Marineland, responsible for replenishing the collection of sea life in the park, as well as juggling duties as the research library curator, coordinating research with universities worldwide, and fulfilling requests for research specimens from other facilities.

One such researcher was Dr. Joseph F. Gennaro. Gennaro was part of a research team in 1956 that discovered radioactive iodine absorbed through a frog's skin would not only collect in the thyroid as expected but also was being synthesized in other organs, most notably the frog's poison glands. When he became an assistant professor at the University of Florida's College of Medicine, he expanded on the idea and began injecting venomous snakes with radioactive iodine and a thyroid-stimulating hormone. The radioiodine circulated through the snake and ended up in the venom glands. The milking of the snake resulted in a highly radioactive venom. When injected into test animals, this radioactive venom could be traced through the test animal's body. This allowed researchers to follow the route and speed of the venom through the test animal and pinpoint how and where venom caused death. This was a new tool in developing antivenoms. Gennaro decided to expand from snakes into other types of venomous creatures to see how other venoms killed. His new expanded experiments included fish and octopuses.

Gennaro contacted Marineland for octopuses. Genaro and Wood became friends, as reflected in correspondence between them that grew more casual in tone as Woody and Joe collaborated. In 1957, Gennaro decided to inject octopuses with a thyroid-stimulating hormone and then add radioiodine to the tank. This required Wood to locate a pair of octopuses of similar size, one to inject and one to act as a control, and then a second similar pair to repeat the experiment.

As Wood struggled to locate four octopus of roughly the same size for the radioiodine experiment, the conversation turned to octopus size and growth. Wood showed Gennaro a newspaper clipping he had found among Marineland's research files on octopuses. "The Facts About Florida" by Frank H. Spalding was an illustration done in the style of the "Ripley's Believe It or Not" strips and featured a small piece on Black Jack Pershing's early career and a drawing of an octopus with the text:

> In 1897, portions of an octopus, said to have been more massive than any other before seen, were washed up on the beach at St. Augustine. Prof. Verrill, of Yale University, who examined the remains, which alone reputedly weighed over six tons, calculated that the living creature had a girth of 25 feet and tentacles 72 feet in length!

It was undated and unsourced. Wood had tried to find a source but was unsuccessful. Part of the problem was that he assumed it was older than it actually was. Spalding's "The Facts About Florida" was a short-lived feature in the *Miami Daily News* Sunday magazine, circa 1946, and had probably been included in one of the packets of articles sent from one of the news clipping services employed by Marineland. Wood was not overly concerned with the source. What he had honed in on was the fact that the giant octopus report came from St. Augustine.

Wood knew St. Augustine had no ocean access, so the octopus had to have washed ashore on Anastasia Island, the barrier island that was also home to Marineland, meaning the octopus had washed ashore within 10-to-15 miles of Marineland. And adding legitimacy to what was an incredulous size for an octopus (the largest known octopus in the North is *Pacific Octopus dofleini*, which reaches up to 125 pounds), the text mentioned A.E. Verrill, still considered an expert in cephalopods decades after his death.

Gennaro was intrigued by the reports of giant octopuses and encouraged Wood to pursue the story. His suggestion was to contact Gilbert Voss at the University of Miami. Voss had just published an article on giant squids that included a section hypothesizing the existence of an even larger squid, the colossal squid (it would be proven to exist in the 1980s). Voss advised an astonished Wood that he remembered the Smithsonian still had samples of the St. Augustine sea monster sent by Webb. Wood began corresponding with Harold Rehder, curator of the mollusk collection at the Smithsonian, looking for the giant octopus samples.

In 1960, Rehder confirmed that the collection included several pieces of the specimen, still in the original 10-gallon jar, hand-lettered "*Octopus* giganteus verrill.*" More importantly, since Rehder considered the specimen a misidentified whale as the 1897 debate had left the matter, he was willing to consider allowing samples to be removed for testing. Gennaro could use the equipment at the University of Florida to perform a histological examination, looking at the microscopic structure of the cells to determine if the cellular structure was that of a mammal or a cephalopod.

The year 1960 may have started on a high note with the discovery of the Smithsonian's samples, but things started to go downhill as soon as Gennaro arrived at the Smithsonian to obtain the samples for testing. He made several observations immediately. Webb had sent six chunks of the carcass, each roughly in the 8-to-10 pound range. Secondly, a 10-gallon jar of a 70-year-old concoction of formaldehyde, alcohol, and 60 pounds of rotten flesh had a penetrating noxious odor of legendary proportions. Gennaro also discovered the pieces also matched Webb's description regarding the toughness of the mass to cut, noting he went through four scalpel blades to remove two finger-sized pieces. Nor was Gennaro particularly happy with his samples, but that problem was with the original samples sent by Webb. The Smithsonian's specimens were all tough, white, and fibrous. There were no distinguishing features such as suckers, skin, or muscles.

Gennaro brought the samples back to the University of Florida. To his dismay, no cellular material was discernible even when examined under the scanning electron microscope. He hypothesized perhaps the mass had baked too long beneath the Florida sun, and the formaldehyde could not penetrate the tough surface for adequate preservation. Disheartened, Gennaro prepared to write to Wood and tell him that nothing of the original cellular architecture remained. Then inspiration struck. The tissue examination may have proven inconclusive, but perhaps the connective tissue could be examined under polarized light.

In February 1961, Gennaro compared the sample to samples of squid and octopus provided by Wood. The connective tissue oriented in the plane of the section refracted under polarized light and showed up brightly; those that were perpendicular to the section appeared dark. The difference between the squid and octopus samples was easily observed. In the octopus, broad bands of fiber across the plane of the sample were separated by broad bands in a perpendicular direction. In the squid, the bands were narrower and separated by thin bands.

Gennaro immediately ran into trouble with the St. Augustine sample. The sample had fewer bands visible, but the bands were not as clearly perpendicular as those in the octopus. It was apparent that the Florida carcass was not a giant squid, but the sample was not identical to an octopus either. Gennaro declared it was close enough to prove the existence of a gigantic octopus with arms 75-to-100-feet long.

The two continued to work on octopuses, but the collaboration was ending. In 1963, Wood resigned his position at Marineland and moved to Point Mugu, California, to head the Marine Sciences Division of the U.S. Naval Missile Center. The two kept in touch, but Gennaro also moved on, first to the University of Louisville, then to New York University. The planned joint paper on the histological evidence of the gigantic octopus seemed to be a lost cause.

However, as Wood's discovery of the clipping in the Marineland files had already proven, the St. Augustine monster would not fade away quietly. At the end of 1970, Gennaro, now an associate professor of biology at New York University, was approached by the editors of *Natural History*, then still published by the American Museum of Natural History. With the renewed interest in the Loch Ness monster being generated by Loch Ness Phenomena Investigation Bureau, they wondered if Gennaro might write something up for the magazine on the Giant Octopus. Gennaro contacted Wood and they took their earlier attempt at a paper and modified it for a more general audience. Their findings were published in the March 1971 issue as three separate pieces: Wood's history of the initial discovery, Gennaro's histological analysis, and Wood's notes on Bahamian scutes. The tripartite article had a whimsical tone to it, somewhat atypical of the magazine's usual tone. The introductory note to the article was the first clue:

After a decade of sleuthing, it can be safely said that the gigantic mass of tissue that washed up on the beach at St. Augustine in 1896 was the remains of an octopus that must have measured, from the tip of one tentacle to the tip of the opposite tentacle, 200 feet. Yes, Victoria, 200 feet.

The material itself was accurate, but the flippancy of the piece, coupled with many readers receiving their issue on April 1st, lent itself to the look and feel of an April Fool's joke. The author's biographies did not help. While the other contributor's submitted headshots or pictures taken during fieldwork, Gennaro sent a gag shot of him posed with a skeletal foot protruding from his pant leg, the "peculiar effects of poisonous snakebites." Wood's photo shows him working with a seal, but notes his favorite exercise was watering his bonsai.

Wood, by then on staff at the Naval Undersea Center in San Diego, was—to put it delicately—unhappy with the published article. *The Wall Street Journal* did report on both the giant octopus and the article's tone. Wood was blunt, "I've written a strong letter to the editor and I personally think that Gennaro was irresponsible. I don't think he took this thing seriously."

The editors at *Natural Science* defended the piece, suggesting that the material was open to multiple interpretations, stressing it was not a hoax. The article did nothing to interest science in reopening the case, but it brought the carcass to the attention of cryptozoology enthusiasts. Wood would become a founding member of the now-defunct International Society of Cryptozoology. Gennaro would briefly also be a member but was not interested in cryptids other than the St. Augustine case. He would be less active in organized cryptozoology but was always available to answer questions, suggest research approaches, and occasionally provide a sample of the carcass for testing.

Of all the cryptozoology and Fortean enthusiasts whose interest was piqued by the *Natural Science* article, perhaps the most enthusiastic was a 22-year-old named Gary Mangiacopra. In 1972, he authored a recap of the history of the giant octopus for the International Fortean Organization's *INFO Journal*, the first in a long stream of articles on the St. Augustine monster. Most of these articles languished in relative obscurity, appearing in *Of Sea and Shore,* a magazine geared toward shell collectors, but Mangiacopra became the octopus's most ardent supporter, focusing on recreating the research timeline through correspondence and tracking down newspaper accounts.

Among the St. Augustine Historical Society's archives are myriads of clippings and correspondence from Mangiacopra, including reminders that his devotion to the investigation was not universally appreciated. Clyde Roper, Smithsonian zoologist and A.E. Verrill's heir apparent to the claim of foremost giant squid expert, had been contacted so often by Mangiacopra that Roper refused to respond to his queries. When Mangiacopra had the St. Augustine Historical Society's archivist Jacqueline Bearden make the inquiry instead, Roper recognized who had actually asked and lambasted the archivist for collaborating with him. Even F. G. Wood, in 1974, felt the need to add a cautionary line to his correspondence with Mangiacopra, reminding him the "creature has not been 'proved' to be an octopus."

Several high-profile mentions brought the St. Augustine carcass back into the public eye in 1973. Cousteau recapped the case in *Octopus and Squid, the Soft Intelligence*, including Wood's theory that the carcass had been carried with the current from the Bahamas to St. Augustine, making the Bahamian coast the place to seek live specimens. Gennaro returned to the topic in the March issue of *Argosy*. The brief article only summarizes the work to date, but the circulation of the magazine would be the pinnacle of the Giant Octopus's exposure to the general public.

In 1974, Bernard Heuvelmans released a revised and expanded edition of *Dans le sillage des monstres marins - Le Kraken et le Poulpe Colossal* in which he apologized for excluding *Octopus giganteus* in the first edition, admitting that his respect for Verrill made him assume the case was closed. While the discussions continued in various cryptozoology journals, awareness of the giant octopus continued to ebb and flow among the public. Mangiacopra's first article in *Of Sea and Shore* was reprinted in the July 1976 issue of the more widely distributed *Fate*. Gennaro appeared in an episode of the US television series *Arthur C. Clarke's Mysterious World* discussing the St. Augustine beaching. Wood also appeared, in a separate scene, discussing a possible giant squid attack on a Navy frigate out of San Diego.

It was at this time that the giant octopus began to attract the notice of Michel Raynal. Raynal, with a background in chemistry, physics, and biology, approached the giant octopus question from an analytical perspective. While Mangiacopra continued to collect anecdotal material in old publications, Raynal began questioning the methodology of both the octopus and whale advocates, asking about details a layperson would not consider. At the 2nd annual meeting of the Society for Scientific Exploration, University of Chicago biologist Roy Mackal reassessed Gennaro's test results in the connective tissue. Raynal approached Mackal, a legend in cryptozoology for his work at Loch Ness, and acted as a go-between to obtain samples of the octopus from Gennaro, who by this point possessed the only known pieces, the same ones he had hacked off the Smithsonian's now-missing specimen.

As Mackal tested the samples, Raynal worked with a piece he kept to test on his own. Raynal hypothesized that the St. Augustine carcass was an unknown species, a giant cirrate (finned) octopus, based on Webb's belief he saw dorsal fins and Verrill's subsequent suggestion that such fins made the carcass that of the comparatively primitive cirrate species. Webb had further suggested that the remains had strands he compared to hair, as opposed to merely being frayed connective tissue. The cirrates have a fine fringe of cilia along the tentacles, and Raynal recalled that Wood had mentioned back on his 1956 trip that one of the Bahamian nicknames for the giant octopuses was "Him of the Hairy Hands." Mangiacopra had recently confirmed Wood's thought that the Florida current could push a mass from the Bahamas to the Florida coast. This convergence of theories seemed compelling, and Raynal proposed to change the scientific name of *Octopus giganteus verrill* to *Otoctopus giganteus*, the giant-eared octopus.

Roy Mackal's paper from the Society for Scientific Exploration meeting appeared in the 1986 issue of the journal *Cryptozoology*. His test was based on amino acid residue. After verifying that formalin did not affect the results, he compared the St. Augustine sample against collagen from other marine species provided by Gennaro. A third test compared the amino acid composition to human and bovine collagen in tendons and bone. He determined the sample was almost pure collagen, which had already been established, but this was to be expected, as a bigger creature meant more collagen.

Attached to the report was the iron/copper comparison that Raynal had made arrangements to have tested at the University of Aix-Marseille. That test determines the comparative ratios of copper-based hemocyanin and iron-based hemoglobin, the former being the protein that transports oxygen in the bodies of invertebrates, the latter in a vertebrate. The results were disappointing to Raynal – the ratio was indicative of an invertebrate, but the test on a whale sample, used as a comparison, tested even higher in hemocyanin. He declared the test inconclusive. He was not happy with Mackal's work either, raising questions in the next issue of *Cryptozoology* about results, quality control, and choice of comparative samples. Mackal ignored the insult, having already moved on to writing a book on his expeditions to the Congo in search of the living dinosaur Mokele-mbembe.

The 1990s saw Raynal continue to argue for additional testing to support his theory that the St. Augustine carcass was a giant cirrate octopus while Mangiacopra continued to publish historical material supporting the existence of the giant octopus and other cryptids. Forrest G. Wood died in San Diego in 1992, the same year Mangiacopra received his master's degree in biology from Southern Connecticut State University. After working with Mackal, Mangiacopra's interests had been turning more toward freshwater cryptids such as the Lake Champlain monster, as evidenced in his master's thesis, *Theoretical Population Estimates of the Large Aquatic Animals in Selected Freshwater Lakes of North America.*

The April 1995 edition of *Biological Bulletin* changed the dynamics of the research. A team led by Sydney Pierce of the Department of Zoology at the University of Maryland performed their own electron microscopy and amino acid analyses on both the St. Augustine remains and the smaller mass that washed ashore in 1988, usually referred to as the "Bermuda Blob." Their results again confirmed the carcass was virtually pure collagen. However, they found "neither sample had the biochemical characteristics of invertebrate collagen, nor the collagen fiber arrangement of octopus mantle." They concluded that the Bermuda Blob was from a poikilotherm (i.e., shark) and the Florida sample was from a huge, warm-blooded creature (i.e., a whale) that been dead and in the ocean so long that bacteria had rendered it down to only collagen, the most resistant protein. They concluded that there was "no evidence to support the existence of *Octopus giganteus.*"

The report galvanized the cryptozoology community. The counter-arguments relied on discussion points familiar to the various authors, such as eyewitness accounts, test results, and questions of comparative testing samples. Heuvelmans misspoke as to the condition and number of tentacles on the beach. J. Richard Greenwell, the Secretary of the International Society of Cryptozoology, went so far as to accuse Pierce of having an agenda against the carcasses. Instead of raising doubts, it made the authors appear shrill and overly-defensive. Mangiacopra and Raynal continued to press for additional tests. With the dwindling amount of tissue left, requests were becoming increasingly hard to justify to Gennaro, who was planning his own test series.

In 2003, two years after Heuvelmans's death, an English edition of his revised sea serpent book was released as *The Kraken and the Colossal Octopus: In the Wake of Sea-monsters.* In the aftermath of the *Biological Bulletin* debunking, Heuvelman had gone back and included revisions and addendums for the English release, most heavily derived from the work of Mangiacopra and Raynal.

In July of 2003, a large mass of tissue was discovered on Pinuno Beach in Los Muermos, Chile. Cryptozoologists had a sudden sense of déjà vu. The mass, named "the Chilean blob," weighed 14 tons and measured 39 feet across. It looked like the St. Augustine remains, was of a comparative size, had similar coloring, and even had strands trailing off that resembled tentacles.

Sydney Pierce, now of the Department of Biology at the University of South Florida in Tampa, immediately received a sample taken from the blob in situ within days of the discovery. Some of the tissue samples were preserved in ethanol, and some were frozen. The material was shipped to Tampa by overnight express. Whether Pierce had heard the complaints from the open forum after his last paper, or he was just ready to resolve the matter once and for all, Piece pulled out all the stops. His team tested the sample, as well as samples from other "blobs": St. Augustine, the two Bermuda blobs, and others found at Nantucket Island, Massachusetts, and the Tasmanian West Coast. He had a second independent analysis of the Chilean Blob carried out in Auckland, New Zealand, by Carlos Olavarria of the School of Biological Sciences, University of Auckland. The results were devastating to the Giant Octopus supporters.

The microscopic anatomy of all the carcasses proved to be virtually pure collagen fiber in cross-hatched layers, precisely the same as the collagen fiber of humpback whale blubber used for comparison. Alternately, octopus and squid consist of muscle fibers with few collagen fibers. The amino acid compositions of all the globs were similar and also indicative of collagen. They extracted DNA from the Chilean and Nantucket samples (the others were too badly damaged by their preservation methods).

Their conclusion, published in the June 2004 *Biological Bulletin,* was that the DNA sequence was 100% identical to the *nad2* gene found in the sperm whale (*Physeter catadon*). It was irrefutable evidence that the Chilean blob was the badly decomposed remains of a sperm whale. With the microscopic anatomy and biochemical similarities, there was no doubt that all the remains, St. Augustine included, were derived from the same source—they were all the decomposed remains of the blubber layer of one of the great whales. Even St. Augustine's most ardent supporter, Michel Raynal, was convinced. The spermaceti tank of a sperm whale, when emptied of the up to 500 gallons of waxy liquid that gives the tank its name, resembles a large, empty collagen sack. When taken out of the whale and out of context, the large, empty sack may resemble an octopus, but it is not, unfortunately.

Gennaro remains a vigilant custodian of the few remaining pieces of the

"St. Augustine Monster." He continues to quietly analyze the samples he took from the Smithsonian in 1960, using the latest techniques as time and opportunity permit. He remains convinced the carcass is not a cetacean.

However, it now appears that when Herbert Colee and Dunham Coxetter rushed to Dr. DeWitt Webb's St. Augustine home on that Sunday afternoon in November 1896 to tell him a whale had beached on South Beach, they were right all along.

The St. Augustine Octopus was killed by science, but the monstrous carcass has survived, both in the "cut and paste, verification optional" world of the internet and the "sensationalism over accuracy" programming of cable television. If nothing else, the St. Augustine Sea Monster is proof you can't kill a good story.

# CHAPTER 7
## Atlantic Encounters

As mentioned in the Introduction, sea serpents have no sense of geography. The most notable of these sightings along the East Coast of Florida takes place around the mouth of the St. Johns River, 20 miles from the Georgia border, and follow the coast both north and south. Although Heuvelmans believed sea serpents of the warm climes differ from the northern cold-water serpents, it appears that charting 50-to-100-foot serpent sightings along the coast would create an unbroken line from the Caribbean into the Canadian Maritimes.

### The Schooner *Eagle*

The schooner *Eagle* arrived in Charleston, North Carolina, on Saturday, March 27, 1830, from Turtle River, Georgia, via Tybee Island, hauling cotton and merchandise. As the crew unloaded the cargo, word got back to the *Charleston Courier* that the ship had encountered a sea monster.

The *Courier* approached the *Eagle's* captain, Joshua Delano, a familiar face at the port. Captain Delano had been running cargo up and the down for coast for years, although since a July 1827 wreck on the Cape Lookout Shoals, he had limited himself to short runs to and from Georgia.

Captain Delano agreed to tell the newspaper about his encounter. On Monday morning, the *Eagle* had departed Turtle River. About 10 a.m., the ship was a mile inside Saint Simons Bay heading toward the open sea. The crew observed a large creature resembling a partially submerged alligator about 900 feet away. It seemed to be paralleling the *Eagle*. Captain Delano was concerned as the creature was gradually getting closer. He loaded a musket and directed the help to change course and get within 100 feet of it. The ship got within 75 feet when the captain fired the musket at the base of the monster's head.

The musket got a reaction, just not the one he expected. Instead of being wounded or fleeing, the serpent charged the ship. At the last moment, it turned and skirted the vessel, striking the ship several times with its tail, shaking the ship. The crew had the opportunity to observe the serpent up close—

perhaps a little more closely than they would have preferred—but it resulted in one of the more detailed descriptions of a regional sea serpent. They agreed that its length was upwards of 70 feet, with the body as large, or larger, than a 60-gallon cask. It was gray in color and shaped like an eel without any visible fins. Apparently, it was covered in scales or plating—the serpent's back was described as "full of 'joints' or 'bunches.'" The description noted a 10-foot-long head and a mouth that resembled an alligator.

The captain also noticed a smaller version of the creature further away, which vanished when the musket discharged, but both serpents were seen together near the north shore, where they finally disappeared. Captain Delano considered himself fortunate for such a short encounter—he was certain that had the serpent been so inclined, it could have done significant damage to the *Eagle*.

Opinion was mixed as to what Delano saw. One later article in a Savannah newspaper noted a whale had been seen several times in St. Simons Sound. Aside from a different length and description, the whale sounded like it could be Captain Delano's monster. The debate continued through the summer when the carcass of a small sperm whale was found on Jekyll Island by campers. That seemed to have finally exhausted interest in the debate.

Dutch zoologist Antoon Cornelis Oudemans believed the story to be a hoax precisely because of the serpent's decision to charge the *Eagle*; behaviorally "it is not the habit of the sea-serpent to attack a ship after being struck by a ball, but to plunge down and disappear." However, the detail of the second, smaller sea serpent may suggest the behavior was logical, assuming that sea serpents have parental instincts.

If Captain Delano comes across as unperturbed by a close encounter with a sea serpent, it may be because he readily admitted it was not his first encounter. In 1826, he had encountered a similar serpent about 20 miles up the coast in Doboy Sound. He had fired three shots at that one but didn't get quite as aggressive a response as the more recent encounter.

## The Schooner *Lucy & Nancy*

The *Lucy & Nancy* docked in Jacksonville, Florida, on February 22, 1849. Launched in 1834 out of Penobscot, Maine, the ship was well-known along the eastern seaboard. It was a typical schooner of the time, two-masted, 72 feet in length, with a register of 101 tons, and it could carry about 150 tons of cargo. She was designed for fast trips with heavy payloads; its crew of just three or four men made her profitable as well. The ship, usually running lumber out of Maine to mid-Atlantic ports, had just completed a trip from New York. What made this mundane trip memorable was that the captain reported they had encountered a sea serpent.

On February 18, 1849, the *Lucy & Nancy* was off the south point of Cumberland Island, which marks the Florida boundary with Georgia. Twelve miles away was the treacherous St. Johns bar at the mouth of the St. Johns River, which leads to Jacksonville. Captain Adams, his crew, and his passengers suddenly saw "an immense sea monster," which the captain took to be a serpent. Captain Adams described the beast as having the head of a snake and having an unnerving tendency to lift its head out of the water to watch the ship, which exposed "a pair of frightful fins or claws several feet in length." The creature was estimated to be 90 feet long and dirty red and brown in color. The creature's neck tapered from the head to the body, where it reached seven feet wide across the back. The serpent sped out beyond the ship and lolled directly in its path. Captain Adams, not wanting a close encounter with the creature, ordered the ship's path changed to avoid it. The youngest crew member had grabbed a harpoon and was preparing to meet the beast's apparent challenge when the ship veered off and avoided the issue. Like the ship, the serpent appeared to be heading to the mouth of the St. Johns.

The captain was David E. Adams, a man with an impeccable reputation both as a mariner and as part-owner of the *Lucy & Nancy*. And because his father and brother were highly regarded ministers in the Knox County region of Maine, Captain Adams's word lent a certain degree of gravitas not customarily offered to sea serpent stories.

Oudemans, in his 1892 book *The Great Sea-Serpent*, observed that the description of the fins/claws was strikingly similar to one of the earliest sea serpent accounts, the 1734 encounter by the "Apostle of Greenland," Hans Egede, somewhere off the coast of Greenland. The account was not written by the missionary but rather by his son Paul in recollections published years later.

Adams' report generated international attention at the time. It had been barely six months since international interest had been generated by an encounter between a sea serpent and the frigate HMS *Dædalus,* in which Captain Peter M'Quhae described a 60-foot-long serpent, 300 miles off the coast of southwestern Africa.

It would not be the *Lucy & Nancy's* only association with sea serpents. On October 9, 1873, the schooner was hauling lumber when she foundered in rough seas. The ship was a total loss, and two seamen drowned. The location of the wreck was off Rockport, site of multiple sightings of New England's most famous cryptid—the Gloucester Sea Serpent.

## The Steamer *William Seabrook*

The steamer *William Seabrook* arrived in Savannah, Georgia on the evening of March 11, 1850. The ship had departed Charleston, South Carolina, and made good time. The vessel had passed the mouth of the Broad River, and as the ship paralleled Hilton Head, they began looking for the Tybee Lighthouse at the mouth of the Savannah River. At about 6 p.m., someone onboard spotted a floating log several hundred yards ahead of the vessel. As the *William Seabrook* got closer, it became apparent that the "log" was actually an animal of some sort, partly submerged in the water. As the steamer came still closer, it raised its head. The passengers and crew described it as being a dark, muddy color, with a head resembling an alligator's raised 10-to-15 feet out of water. When the body also partially surfaced, it "exposed numerous bumps of the size of a hogshead rising out of the water." The estimated length was 140-to-150 feet.

Captain Peleg Blankenship, a "bold, fearless and careful officer, well and favorably known," was not impressed. He ordered the ship to run a mile off course just to circle around the sea serpent, at times getting as close as 20-to-30 feet. The serpent was equally unimpressed and simply submerged. As the ship resumed course, the beast surfaced and swam off to the south.

The Savannah newspapers were elated—they had been following the reports of the Gloucester Sea Serpent up in New England, perhaps with a touch of envy. Captain Blankenship's sighting was exactly what they wanted. Perhaps with tongue in cheek, the *Palmetto Post* in Beaufort, South Carolina, on the north side of the Broad River, notified the Savannah and Charleston newspapers that the serpent had been sighted again, this time in the Whale Branch, a winding, marshy river that runs from the Broad River to the Coosaw River. The serpent was seen and fired upon. The hunters chased it upriver five miles (roughly where today's State Route 21 crosses the river) before it submerged and they lost the trail. The hunting party was undeterred and planned to resume the hunt in the morning with additional boats and a barge with a six-pounder. Whether having a cannon on a barge was such a good idea was not reported, but a week later, the *Palmetto Post* reported that witnesses had spotted the serpent but that it had turned out to be a pod of whales swimming in such as a way as to appear to be the undulating humps of a single beast.

In the spring of 1854, the *William Seabrook* had another encounter. The steamer was heading up the Savannah River when the new captain, Fenn Peck, the crew, and passengers all saw a similar serpent ahead of the vessel. This time there was no sightseeing. The serpent disappeared as quickly as it had appeared.

Amateur botanist Amelia Murray relayed the 1854 sighting in a May 1855 letter to a friend. Murray was aboard the *Isabel* heading from Key West to Cuba, and while it would be easy to dismiss the story as the captain telling tales to entertain passengers, Amelia Murray was no ordinary tourist. Amelia Matilda Murray (1795-1884) was on a year-long exploration of North America from England, where she was a lady-in-waiting to Queen Victoria. Her letters were to an old friend, Annabella Milbanke, later Lady Byron, being briefly married to the poet Lord Byron, and mother of mathematician Ada Lovelace.

The *Isabel's* captain went further, admitting he too had seen the sea serpent in the Savannah River the day after Captain Peck and the *William Seabrook* had spotted it in 1854. Called to the bridge by a lookout, he looked through his telescope at a serpent, as long as the *Isabel*, moving quickly. As he watched, it reared its snake-like body and head high out of the water as high as the funnel of the steamer, looked about for an instant, and then submerged. It wasn't much of an encounter, but Murray considered Captain Rollins a reliable source. And she already believed in sea serpents, based on its inclusion in Scandinavian mythology, which she knew only included source material derived from nature.

Murray may have been touring North America as a botanist, but the farther south she traveled, the more she believed the southern states' use of slaves was a good thing for the enslaved and the owners. When her letters were published in 1856, her advocacy of slavery was a scandal and she forced to out of her position in Victoria's court. The sea serpent report didn't raise an eyebrow.

### The Steam Packet *William Gaston*

The sidewheeler *William Gaston* ran a weekly route from Savannah, Georgia, hopscotching down the coast to Jacksonville and then up the St. Johns River to Palatka, then reversing route back to Savannah, delivering goods, passengers, and the US mail. The passengers arriving in Savannah on January 15, 1853, got more for their money than just passage.

The *William Gaston* left Jacksonville on schedule, on Wednesday the 12th, heading for St. Marys, Georgia. The ship had just passed over the St. Johns Bar and was nearing the outer buoy. The seas were calm, the afternoon skies were clear, and 30 passengers had an unhindered view of a sea serpent.

The *William Gaston* arrived in Savannah on Saturday. The *Savannah*

*Republican* quickly got word of the encounter and interviewed the captain, Thomas E. Shaw. Shaw readily acknowledged specifics. The seas had been perfectly calm, and there, stretched out on the surface, was a large sea monster. Shaw described it as 60 feet long and dark in color, approaching black. The head was as large as a hogshead barrel (a standard hogshead measures 48 inches long and 30 inches wide). The serpent paid little attention to the ship, merely lowering itself slightly deeper into the water. The vessel passed very near it, and there was some call to investigate, but crossing the bar required high tide, so the timing was crucial, and the ship continued along its route.

### The *Saladin* and the Balloon Fish

On March 12, 1870, the Nova Scotian schooner *Saladin* was 130 miles off Wilmington, North Carolina, heading NNW to round Cape Hatteras on her way to New York with a cargo of copper from Haiti. It was 6 a.m., and the *Saladin* was sailing at a leisurely four knots. Captain I. S. Slocomb was at the helm, letting his small crew rest up after several days of rough weather.

When Slocomb spotted an overturned boat five miles off the starboard beam, he called all hands on deck and changed direction to render any assistance needed. As the *Saladin* drew nearer, it turned out not to be a boat. It was an enormous sea monster that dwarfed the *Saladin*. Slocomb estimated it to be 100 feet long: a body 40 feet in length and a tail of 60 feet. But the oddest feature was an immense oval "balloon" of cartilage, 12 feet in height and 40 feet wide. Regular ridges were running the length of the balloon, four inches apart and two inches high, giving the appearance of the rope netting used to contain a hot air balloon. The balloon, Slocomb hypothesized, was filled with air and probably a way for the massive creature to submerge and rise as desired, allowing it to remain on the surface and wreak terror on potential meals such as fish, dolphins, and sailing ships.

On each brownish-colored side were two large fins, each five feet long. About 10 feet from the dome were two eyes, one on either side of a large horn. From this point, the fish sloped down a massive tail. The monster was traveling at two knots per hour, and the captain decided to let the beast be, fearful that any provocation would result in the ship's destruction. He carefully and quickly sailed away,

Upon arriving in New York, Captain Slocomb dutifully reported the encounter to the *New York Herald*. The *Herald* was dubious but wrote an account so flowery and over the top that it was obviously meant to be sarcasm. Instead, the story went international. Inexplicably, it was picked up in 1896, 26 years later, as a newspaper filler piece. The story ran numerous times, essentially unchanged, although it tended to be truncated. The problem is that most versions only mention the longitude and latitude of the encounter, and the degree numbers vary from one newspaper to another, placing the *Saladin* as far south as Daytona Beach. Even Heuvelmans got confused, indexing the incident as a Caribbean sighting in his book.

But Heuvelmans was not confused about the Balloon Fish's identity. It was apparent to him that no one recognized it because it was floating belly-up in the water. It was a dead "fin whale" or "rorqual" (*Balaenoptera physalus*), a baleen whale and the second-largest species on Earth after the blue whale. Heuvelmans based his identification on the "balloon netting," which he recognized as ventral grooves, the series of pleats that run from the chin of a baleen whale to the navel that form a pouch that expands to allow massive amounts of water and prey to enter, be filtered and expelled. Heuvelmans was not surprised at the misidentification. In 1870, Svend Foyn had only just patented the modern cannon launched grenade harpoon, so the sheer size of baleen whales was still protecting them from being slaughtered by whalers, making them less familiar.

Heuvelmans theorized the whale was starting to decompose, and the ventral pouch was collecting gases, creating the "balloon." A rorqual explained the size and coloration. The "horn and eyes" on the torso, he continued, were the penis, partly distended by the gas build-up, and the "eyes" were, as Heuvelmans delicately phrased it, "the natural orifices."

Heuvelmans neglects to mention the two flippers per side, but that is also explainable as a fin whale. The species has dark, oval-shaped areas of pigment called "flipper shadows" that extend beyond and behind the pectoral fins, which to a cautious captain taking observations from a safe distance, could appear as a second pair below the waterline.

The whale was floating in the Gulf Stream, which is typically two knots, the same speed Slocomb reported the sea monster was traveling. With fins pushing against the current, it could appear the whale was still alive. Dead or alive, Captain Slocomb's decision not to harass or attack the beast was prudent. Should a spear or bullet pierce the pouch, the balloon's pop would have been significantly more visceral than a hot air balloon's.

### Mr. Tuttle's Octopus
On Saturday, February 20, 1897, Theodore A. Tuttle and Daniel S. Gilhuly had settled into their rooms in the Hotel Biscayne in Miami. "Judge" Tuttle

was a regular winter visitor in Miami and was already a popular snowbird, a supporter of the fledgling local art scene, and popular with the ladies. Gilhuly had turned 50 the week earlier, and he and Tuttle had been celebrating with an extended fishing trip at resorts along the Florida East Coast Railway. They had arrived after a week of fishing in Rockledge where an old friend, Albert C. Hendrick, was vacationing.

Flagler's Florida East Coast Railway had arrived in Miami, which had recently incorporated, and the area was already beginning to see the arrival of land speculators. Undoubtedly Tuttle was not averse to exploring real estate investments while in town.

The 66-year-old Tuttle and the birthday boy decided they would indulge in some afternoon fishing. The two rented a boat and headed out into Biscayne Bay. They were told that others had been having luck off at Bear Cut, the channel on between Virginia Key and Key Biscayne. It was about three miles from the docks, so they did not plan to stay long.

Gilhuly and Tuttle spent a half-hour with tremendous success before the fish suddenly stopped biting. The sun was starting to set, and they began thinking of heading back. Gilhuly was fishing off the port stern with his back to Tuttle, who was casting off the starboard bow. Suddenly, Gilhuly heard a cry from Tuttle. He turned to see a large, crimson-colored snake wrapped around Tuttle's neck and trailing off into the water. Gilhuly grabbed a hatchet and hacked at the snake until it released his friend. Then five more "snakes" popped out of the water. It was then they realized the snakes were the tentacles of an octopus.

What happened next depends on which version of the newspaper article you read. The stories might as well be of two different encounters. The dateline of the article quickly tells the source. In 1897, Dade County encompassed the territories that make up modern Miami-Dade, Broward, Palm Beach, and Martin counties. There were only two newspapers in Dade County at the time. Still sparsely populated, the comparatively more populous area around Palm Beach had outvoted the Miamians and moved the county seat to Juno, near present-day Juno Beach, not returning to Miami in 1899. However, Juno suffered a major fire in 1894, and rather than let their rivals in Miami reclaim the county seat, they moved the seat to Palm Beach during the rebuild. The dateline Palm Beach indicates the story's source was the weekly *The Tropical Sun*, also temporarily in Palm Beach, but considered the "official" newspaper for Dade County by virtue of being in the county seat. It would also be receiving the story second-hand. The second newspaper, *The Miami Metropolis*, was also a weekly that covered the Miami vicinity.

Neither newspaper actually published a story about the encounter. Since the Tuttle encounter took place on a Saturday afternoon, and *The Tropical Sun* published on Thursdays and *The Miami Metropolis* on Fridays, it would be old news by the next week's issues. Instead, both *The Tropical Sun* and *The Miami Metropolis* filed reports with the Associated Press on Sunday with the newspaper reports first appearing across the east coast on Monday. The *Metropolis* version is the more accurate, considering Tuttle and Gilhuly were local guests; *The Tropical Sun* version is obviously from secondhand sources, neglecting details and elaborating on others.

In the *Miami Metropolis* version, carrying the Miami dateline, Tuttle had hooked the octopus, and Gilhuly grabbed a hatchet and hacked at the tentacle until it released his friend. After that, Tuttle fought off two more tentacles with an oar while Gilhuly attacked other arms with the hatchet, concerned the beast might be trying to scuttle the boat. Already the gunwale on the stern was at water level with water starting to pour in. Gilhuly was able to sever the tentacles, allowing the boat to right itself and avoid swamping. The damage to those appendages was enough to drive the creature off; the tentacles Tuttle was battling also disappeared as the boat righted.

In *The Tropical Sun* version with the dateline Palm Beach, Gilhuly wasn't even in the boat—it was Theodore Tuttle and his hatchet, chopping off tentacles in a battle of life and death. In this version, the octopus attempts to climb into the boat and Tuttle throws the hatchet at the head, driving it back into the water. He then grabs a fishing knife to sever the remaining limbs. This version neglects to mention that Miami was the location, leaving editors running the piece to assume it took place off Palm Beach, further adding to the confusion.

In both versions, the beast vanquished, he/they rowed the boat ashore and was/were helped back to the hotel. Neither man received life-threatening injuries, but both were exhausted. *The Tropical Sun* version had Tuttle suffering, in varying degrees, depending on the subsequent retelling, inflammation and skin irritation where the tentacles had made contact with bare skin.

Tuttle and Gilhuly were former high-ranking officials from New Haven, Connecticut. Tuttle had recently retired after 20 years as the City Tax Collector, and Gilhuly was an Alderman and former police commissioner. Their friend Albert Hendricks in Rockledge was the retired New Haven Fire Chief. As a result, the story was newsworthy in Connecticut. The version of the story that ran in New Haven is the Tuttle-only version from *The Tropical Sun*, which was also the more widely distributed story of the two.

Assuming that high-ranking, municipal officials have no reason to concoct a story of an encounter with an octopus, the question becomes: what could have attacked the small boat? Biscayne Bay's most common cephalopods are the Atlantic pygmy octopus *(Octopus joubini)* with a mantle under two-inches long with arms up to four-inches long; the Caribbean reef octopus (*Octopus briareus*), whose mantle that rarely exceeds five inches and arms that don't exceed 24 inches; and the Caribbean reef squid (*Sepioteuthis sepioidea*), which isn't much bigger, with a mantle less than eight inches. There are also common octopuses (*Octopus vulgaris*) in the bay, which are larger at nine inches in mantle length and arms up to three-feet long, but even a creature of that size would find grasping a boat difficult, let alone sinking it.

The Tuttle account is not an isolated case. Explorer F. A. Mitchell-Hedges describes a similar encounter at the mouth of the North Stann Creek River in Belize in the 1930s, and in the 1950s French journalist François Poli encountered shark fishermen in Cuba who decided that whatever had dragged three metal buoys so deep beneath the surface that they collapsed from the pressure was probably not an octopus. Nobody determined what the creature was, but Poli was intrigued by one fisherman who believed it was the same beast he had encountered on a desolate section of the Mexican coast, a particularly aggressive monster with a yellow-striped cylindrical body and tentacles similar, but not identical, to those of an octopus.

This latter report from Mexico may offer up a less common species as an option—the legendary giant squid (*Architeuthis dux*). Clyde Roper, the Smithsonian's zoologist emeritus and the world's foremost authority on *Architeuthis*, cataloged 16 sightings and specimens of giant squid from the Straits of Florida and Bahamas just in the period 1950-2011. Two of those were found near Fowey Rocks Light, placing two of the creatures on the south side of Key Biscayne, less than five miles from Bear Cut.

But there is also a cryptid that could offer an alternative to the giant squid theory. The concept of a massive tentacled beast emerging from the sea and attempting to drag a boat underwater is a familiar one to the fishing communities of the Bahama Islands, a scant 50 miles to the east. For centuries Bahamian sailors have known and feared such encounters. The Bahamian term for the creature is the Giant Scuttle (Chapter 5).

**The Key West Serpent That Wasn't**
The Royal Yacht Squadron is one of the most exclusive yacht clubs in the world, and even by the lofty standards of the Squadron, the *Valhalla* was one

of the most beautiful large yachts ever built. Lord Crawford's *Valhalla* was a 245-foot, three-masted, full-rigged ship, which he used on numerous occasions for science expeditions. Two seasoned British zoologists, Michael J. Nicoll and E. G. B. Meade-Waldo, were aboard for a specimen collection trip. On December 7, 1905, at 10:15 a.m., Nicoll spotted an unusual fin 300 feet away. The fin was roughly rectangular, about six feet long, two feet high, and "dark seaweed-brown, somewhat crinkled at the edge." Meade-Waldo, using high-powered binoculars, watched as a head on a long neck rose in front of the frilled fin. He later described the creature's neck as "about the thickness of a slight man's body, and from seven to eight feet was out of the water; head and neck were all about the same thickness." The head had a turtle-like appearance, and the creature moved its head and neck from side to side. Beneath the water, the scientist could indistinctly see a very large brownish-black patch, but could not make out the specific shape of the creature's body. Because the *Valhalla* was under sail, she couldn't come about for a closer look, and after several minutes, the ship had passed from visual range. At 2:00 a.m. on December 8, three crewmembers saw what they assumed to be the same animal, this time almost entirely submerged but splashing as it easily passed the ship, which was cruising at eight knots. The story of the *Valhalla* sea serpent is notable for its eyewitnesses being noted naturalists.

It was also apparently a magical ship, capable of being in two places at once. There are publications, borrowing from each other, that identify this encounter as occurring off Key West. But there are far more books that correctly note the *Valhalla* sighting as occurring 15 miles east of the mouth of Brazil's Paraíba do Norte River, which empties into the Atlantic Ocean north of João Pessoa, the state capital.

The prima facie source of the Key West mislocation was James B. Sweeney in his legendarily inaccurate *Sea Monsters: A Collection of Eyewitness Accounts* (1977). He moved the *Valhalla* serpent sighting from Brazil to Key West and then confused the description of the ship with another ship, the *Happy Warrior*. It may have been sloppy research—the *Valhalla* had been sailing up the east coast, pausing in San Juan and then Key West eight months earlier (in April 1905), on her way to New York for the start of a transatlantic yacht race.

On the other hand, Sweeney also concocted a still-repeated tale about German U-boat 85, captured and scuttled off Scotland in World War I. The captain explained the U-boat was cruising on the surface because of a sea monster attack that had done so much damage they couldn't submerge. The book, published and sold as a Young Adult title, continues to be used as a reference tool by less discerning authors and journalists.

## SS *Craigsmere*

After World War I ended, the Merchant Marine steamer *Craigsmere* was used for hauling West Virginia coal from the port in Norfolk, Virginia, to markets up and down the Atlantic seaboard, and beyond, as far south as Colón, Panama. On a routine trip to Nuevitas, Cuba, the *Craigsmere* was traveling between Fort Lauderdale and Miami in sight of the shore (less than three miles), when the captain, the helmsman, and several crew members saw a sea serpent. They described it as long, with multiple dorsal fins and a partially submerged head on a long neck.

Bernard Heuvelmans had been working on classifying categories of aquatic cryptids and grouped both the *Mauretania* and *Craigsmere* sightings as examples of a "many-finned" sea serpent. The decision is an odd one, as Heuvelmans seems to have based the inclusion exclusively on the reports of multiple dorsal fins while disregarding the discrepancies. Under Heuvelmans's categories, many-finned sea serpents, except for the two Floridian sightings and a similar report by the *St. Olaf* in the Gulf of Mexico, were exclusively reported in the South China Sea.

Heuvelmans theorized that such animals were the descendants of ancient whales, fossils of which had been excavated with bony plates. The many-finned sea serpent, he decided, was covered with large scales or bony plates that formed a segmented armor. This armor included lateral projections that look like forward-pointing fins, 4-to-12 per creature. Because the armor restricted movement, Heuvelmans concluded the creature swam with vertical undulations like a snake. But to turn, it had to roll over onto one side, making the other side's lateral armor visible above the water to be mistaken for dorsal fins facing in the wrong direction, which none of the witnesses mentioned.

In addition to the three Floridian accounts being geographic outliers, only the captain of the *Mauretania* mentioned multiple fins—the primary witnesses aboard his ship did not. All *Mauretania* witnesses agreed the serpent was jet black, while Heuvelmans's categories specify that the color was limited to brown or greenish-gray with dirty-yellow speckles.

The primary problem with the *Craigsmere* sighting is the provenance. Heuvelman's source for the account was the records of Fortean icon Ivan T. Sanderson. Sanderson's source was Charles M. Blackford III. Blackford served in the Navy in World War I and the Coast Guard in World War II. In between, 1919-1931, he served in the Merchant Marines. In 1947, he wrote to Sanderson several times, relating reports of strange marine animals from his former shipmates in the Merchant Marines. So the *Craigmore* account was relayed 27 years later by someone who heard it second-hand. That is not a particularly stringent standard for inclusion.

Loren Coleman and Patrick Huyghe attempted a new classification system in *The Field Guide to Lake Monsters, Sea Serpents, and Other Mystery Denizens of the Deep*. They renamed the category "Great Sea Centipede," the name coined by Roman rhetorician Ælian Tacticus in his *On the Nature of Animals* (200 AD); it's believed to be the earliest description of the creature. They also warn this creature is the least likely to be discovered because of the lack of potential fossil progenitors. Heuvelmans's theory had been subsequently dismissed—the bony plates found with the whale fossils he used to build his theory are now known to come from other species that died near the whale. Thus, there is no evidence that an ancient whale with segmented armor ever existed. The inclusion of the category by Coleman and Huyghe, regardless of the creature's lineage, is based on the sheer volume of reports in the South China Sea.

### SS *Santa Clara*

On Tuesday, Dec. 30, 1947, the United States Hydrograph Department received the following wireless message from John Fordan, captain of the Grace Line steamship *Santa Clara*, which was 118 miles east of Cape Lookout, North Carolina, bound for Cartagena, Colombia:

> LAT. 34.34 N LONG 74.07 W 1700 GCT STRUCK MARINE MONSTER EITHER KILLING IT OR BADLY WOUNDING IT PERIOD ESTIMATED LENGTH 45 FEET WITH EEL LIKE HEAD AND BODY APPROXIMATELY THREE FEET IN DIAMETER PERIOD LAST SEEN THRASHING IN LARGE AREA OF BLOODY WATER AND FOAM SIGHTED BY WM. HUMPHREYS CHIEF OFFICER AND JOHN AXELSON THIRD OFFICER.

Captain Fordan was following the rules. His ship had struck an object at sea. He may have been overly cautious in doing so, but the *Santa Clara* was a brand new vessel, less than four months old, and only one of three cargo-passenger combinations that had been built with specific modifications for Grace Lines using the uncompleted hulls of wartime C2 fast turbine freighters. So Fordan was going to document any irregularity in the vessel, even if that irregularity was a collision with a sea monster and not really an issue with the ship herself.

The US Hydrograph staff, not considering a sea monster to be a navigation hazard, passed the wireless message on to the Coast Guard, who let it leak to the press. By the next day, it was in *The New York Times* and on the Associated Press newswire. There was so much interest that the Associated Press wired Captain Fordan asking for additional details. By that evening, Captain Fordan had agreed and radioed a more detailed report to the AP in New York:

On Dec. 30, 1947, the Grace Line steamer Santa Clara was cleaving through sunlit calm blue seas 118 miles due east of·Cape Lookout, en route from New York to Cartagena.

The Santa Clara had just crossed the Gulf stream when William Humphreys, chief mate, John Rigney, navigating officer, and John Axelson, third mate, assembled on the starboard wing of the bridge to take the noon sight at 11:55 a.m.

Suddenly, John Axelson, the third mate, saw a snake-like head rear out of the sea about 30 feet off the starboard bow of the vessel. His exclamation of amazement directed the attention of the two other mates to the sea monster, and the three watched it unbelievingly as it came abeam of the bridge where they stood, and it was then left astern.

The creature's head appeared to be about two and one-half feet across, 2 feet thick, and 5 feet long. The cylindrically shaped body was about 3 feet thick and the neck about one and a half feet in diameter. As the monster came abeam of the bridge, it was observed that the water around the monster, over an area of 30 or 40 square feet, was stained red. The visible part of the body was about 35 feet long.

It was assumed that the color of the water was due to the creature's blood and that the stem of the ship had cut the monster in two, but as there was no observer on the other side of the vessel there was no way of estimating what length of body might have been left on the other side.

From the time the monster was first sighted until it disappeared in the distance astern, it was thrashing about as though in agony. The monster's skin was dark brown, slick and smooth.

There were no fins, hair, or protuberances on the head or neck or any visible parts of the body.

One day out of New York, and barely 400 miles into its journey to Colombia and Venezuela, the *Santa Clara* managed to avoid much of the ensuing media coverage. It was just as well. In their absence, reports had painted the crew as liars, drunks, publicity hounds, or incapable of rudimentary marine creature identification. United Press and *The New York Times* dismissed the creature as an oarfish (*Regalecus glesne*). Christopher Coates, the curator aquarist of the Aquarium of the New York Zoological Society, believed they hit something, which he thought was an oarfish, although he vaguely noted several details didn't match. The unspecified details included shape, the color, and the lack of fins. Oarfish are ribbon-shaped, silvery-blue in color, and have a distinctive pinkish-to-bright red dorsal fin that looks like a spiky headdress extending the length of the fish's body.

On January 1, when a 60-foot long, 35-ton whale carcass found 50 miles off Cape Henry in Virginia was being hauled in by tugboat to a rendering company, the Associated Press ran with the explanation that it must have been the monster that collided with the *Santa Clara*. When a second whale carcass found near the Norfolk Naval Station the next day—this one 61-feet long and weighing 45 tons—the towing company noted that migrating whales travel in pairs, so obviously the ship had struck both whales in the same incident. That pushed the theory into absurdity—essentially declaring the command staff of the *Santa Clara* couldn't tell the difference between hitting 80 combined tons of whale versus a 45-foot long snake-like creature. Additionally, the damage on the whales was reported as superficial in appearance, far less than expected after striking a fast-moving vessel weighing nearly 9,000 tons fully laden. The damage was insufficient for the amount of blood witnessed aboard the ship.

By the time the *Santa Clara* returned to New York on January 14, interest had subsided, but journalists were still waiting to interview the command staff. They had nothing to add to the account, but it was not for lack of trying by the media. The reporters were given access to the ship's log—the collision was entered in red ink, standard procedure for the mandatory entry required when physical contact is made with a floating object or vessel.

If there was any remaining fallout for Captain Fordan and the three senior deck officers who witnessed the sea monster, it vanished the next month when Fordan and his crew were able to reverse course 25 miles with such accuracy that they were able to make a miraculous rescue of the ship's carpenter who had fallen overboard.

The *Santa Clara's* encounter remains a fixture in cryptozoology as one of the most specific descriptions of a sea serpent available, made by four seasoned, professional mariners. Ivan Sanderson was one of the reporters who interviewed the four witnesses, and he was impressed by the consistency among the accounts given by the four men. As far as he was concerned, the *Santa Clara* "had rammed a very large animal of unknown and unidentified type off the eastern North American coast; period!"

## New Smyrna Beach

In February 1959, Robert S. Browning of New Smyrna Beach wrote a letter to Marineland on Anastasia Island. His wife was terminally ill, and the 65-year-old Browning was thinking of his own mortality. He had seen a sea serpent in 1948 and it still bothered him. Although no one believed him, he felt there should be a permanent record of the sighting somewhere, so he was putting it down on paper.

Browning had visited Marineland recently, and it had reminded him again of his own sighting. The head of research, Dr. F. G. Wood, received Browning's letter; his reply was carefully noncommittal. Browning was not describing a giant octopus or a giant squid, which were Wood's interests. Wood would later be responsible for rediscovering the St. Augustine Giant Octopus. Browning's letter was carefully filed in a folder on sea monster reports from newspapers and letter reports sent to him about (usually misidentified) sea monsters.

Browning lived on Atlantic Boulevard in New Smyrna Beach. He lived less than 500 feet from the shore and was no stranger to the ocean. He explained in his letter that he lived in this house, first as a summer home and then fulltime after inheriting it from his parents, for 20 years. His childhood was spent in Peconic on the far end of Long Island. His point was that although he knew coastal waters well, he still didn't know what he saw in 1948.

The year 1948 was already a memorable one. The town he lived in, Coronado Beach, had been annexed by New Smyrna and the two were now known as New Smyrna Beach. In the fall, a school of small whales had beached near the house. Looking at the carcasses, he saw a strange object in the water about 150 yards offshore. Wondering if more whales were going to come ashore, he grabbed his binoculars and went to the beach to get a better look at what was out there.

Through [the binoculars], I could see what was the body of some creature, it appeared flabby fat and was coal black on the top. As it turned very slowly it would heave up on a swell and part of its undersides would come into view which disclosed that the belly part was of a sulphur yellow color and between the black and yellow was a series of irregular black blotches which were large near the black and grew progressively smaller towards the yellow. There was no sign of any fins or other appendages. As I watched the creature seemed to move its neck and head back and forth as if chasing something. This was done with great power for it threw a mass of spray six or eight feet into the air fifteen or more feet ahead of the body part which remained stationary. This occurred several times and I must have watched for close to twenty minutes. At this time a shrimp boat which was a couple of miles away began to draw nearer. I do not know whether this was a signal for the creature to disappear or whether he grew tired of what he was doing but he made up his mind to go and this is the part that really amazed me and caused the incident to make such an impression on me for as he started slowly to sink (he did not rush off) he raised from the water behind him a long fairly slender black tail which must have been twenty to twenty five feet long and which curved in a most graceful arc and slowly following the line of the arc descended into the water as the rest of the creature disappeared. This tail had no fins nor did it have flippers and the end of it did not come to a point but was cut off square.

Browning estimated the creature was 60-80 feet in length, and that when it raised its tail, the cleared ten feet above the water. There is no question that Browning's estimates are probably more reliable than most sea serpent witnesses. He had a degree in chemistry, and in World War I, he was a naval aviator, tasked with reconnaissance (Browning was a Quaker and exempt from combat). In World War II, he was the first commander of the USS *LST-388*, an amphibious landing craft in the North African campaign. At the time of the sighting, he had just retired as a Lieutenant Commander in the US Navy Reserve.

The question, aside from what the creature was, might be whether the creature's presence had driven the whales ashore, or whether the creature was feeding on crabs and other scavengers attracted by the decomposing whales.

## The Giant Juno Worm

The Juno Ledge is a popular diving destination. The two-mile long reef starts at 68-feet below the surface and extends 20-feet down to the sand to a depth of 92-feet. Running north to south, the ledge, with caves and overhangs, offers glimpses of loggerheads, green moray eels, and goliath groupers among the soft corals.

Jay Garbose, a retired attorney living in North Palm Beach, was enjoying his retirement as an underwater videographer. Garbose was diving with a charter boat a mile off Juno Beach at the Juno Ledge on the morning of April 14, 2007. He was exploring around the seafloor when he saw something he didn't recognize. As it turned out, neither did anyone else.

Garbose's six-minute video shows a worm-like creature, gray and flat, able to contract and stretch "like taffy" to seven-to-ten-feet long. He only stopped filming because he needed to return to the surface. He asked other divers to help identify the creature, but they didn't recognize it either, only suggesting it might be some form of a sea cucumber. The only native sea cucumber Garbose could find big enough to be a candidate was the tiger's tail sea cucumber (*Holothuria thomasi*), which it really didn't match well; although it can reach six-to-seven-feet long, it is brown and mottled with patches golden brown and white.

Garbose put the video up on his website and identified the creature as a sea cucumber but also contacted David Pawson at the Smithsonian's Department of Invertebrate Zoology, who he had worked with on a previous underwater documentary. Pawson sent the video to Jon Norenburg, the Smithsonian's expert on marine invertebrates. Norenburg contacted Garbose and told him that his footage was of a nemertean worm, also known as a ribbon worm, some species of which can grow to over 100-feet long, though most can be measured in inches.

Norenburg also informed Garbose that he couldn't determine the family of the worm or tell if it is a new species or not because identifying anatomical details were not visible on the video. The Smithsonian would need additional foot at a closer range or a captured specimen to identify the worm accurately.

Garbose updated the information for the video, which was rapidly becoming a news story, first among diving sites, then with news aggregators. By early May, much to the Smithsonian's annoyance, local television news was declaring that Garbose's "sea serpent" could not even be identified by the Smithsonian.

Oddly, the nemertean worm generated minimal interest in cryptozoological circles, in spite of being mysterious, unidentified, and touted as a new species. Apparently, a ribbon worm cryptid is just not as sexy as a Bigfoot. The cryptid didn't even get a nickname until 2011 when Scott Marlowe's *The Cryptid Creatures of Florida* dubbed the beast "The Giant Juno Worm." The Smithsonian still won't name it until they get a better look.

# CHAPTER 8
## Swamp Serpents of the Treasure Coast

The coastal areas of Martin, St. Lucie, and Indian River Counties are collectively known as the Treasure Coast (some definitions include Palm Beach County). The term, a reference to a Spanish treasure fleet lost off the coast in a 1715 hurricane, was coined in the early 1960s as a way to distinguish itself from Miami and its surrounding Gold Coast in the endless battle for tourist dollars.

The coast is shielded from the Atlantic Ocean by the Intracoastal Waterway, narrow sandbars and barrier islands, creating shallow lagoons, rivers, and bays, including the Indian River Lagoon system, one of the most diverse estuaries in the country. Before humans declared war on the ecosystem in the name of growth, the mainland was part of the Everglades swamps that enclosed Lake Okeechobee. It has a history of oversized serpents in the swamplands.

There have always been reports of giant snakes in Florida—most accounts of river monsters tend to be described as giant snakes. But the stories from the Treasure Coast appear to describe a sizeable euryhalinic serpent that is equally at home on land or in the water, capable of residing in fresh, salt, or brackish water. The region includes the headwaters of the St. Johns River, which is home to a variety of monsters as well.

**Juno Ridge**
The earliest recorded sighting of a serpent along the Treasure Coast took place because of an elopement. Emily Lagow Bell was an early pioneer of the Indian River region and part of a family that helped settle Fort Pierce. Bell wrote a memoir of her time as a settler in the wilds of 1880s coastal Florida as a record for her descendants, not as local history. Instead, it became a primary source of the life of the Florida settler experience after the Civil War.

The cryptid encounter took place in May 1881, when Emily's sister Selene unexpectedly eloped with Abner Wilder. Selene's mother, Lucinda, didn't like this development at all—Abner Wilder was a total stranger, the brother-in-law of a neighboring homesteader, and even more distressing, Selene was 15. Lucinda needed to tell her husband, Alfred, who was off at a hunting camp with Emily's husband, James. By the time Lucinda and Emily loaded the children into a boat and sailed four hours south along the Indian River to reach their hunting camp with the news, it was too late to cross the Jupiter Narrows and intercept the couple, presumably heading for Abner Wilder's home on Lake Worth. So they crossed the narrows in the morning, leaving Emily with the children, while Emily's husband and Selene's distraught parents proceeded by foot the final eight miles to Lake Worth.

Emily recalled her mother's encounter. Lucinda had fallen behind the men as they neared Lake Worth, near what today is called Juno Ridge. She noticed the bushes moving and heard branches snapping. Looking around, Lucinda spotted something slowly dragging itself through the underbrush about 20 feet away. She beckoned Alfred back, and they discovered a sea serpent crawling toward the sea. Emily's mother noted it was "green and black and a yellowish mingled colors." Noticing the humans, it raised its head up three or four feet; Emily's mother thought the face had a human-like quality. The Lagows calculated the serpent's length at about 30-feet long and a width similar to that of a small nail keg (10-to-12 inches).

Arriving in Lake Worth, they ran into settlers David Brown and Charlie Moore, who informed them that Selene and Abner were already married. Their whereabouts were unknown, although they pointed out that Abner's mother lived on the east side of the lake on Palm Beach Island. Since they were too late, the Lagows decided to give up their pursuit. Apparently, Charlie Moore neglected to mention that Abner Wilder's mother had remarried and was now Mrs. Charlie Moore and that his stepson owned a homestead next door to his own.

When the Lagows mentioned their encounter with the serpent, neither man was surprised. The Lake Worthians said it was seen about twice a year, adding that the Indians were superstitious and wouldn't capture it, let alone hunt it.

Bell mentions another sighting in 1927, the year before her memoir was published. She noted this new serpent had been spotted twice on Surfside Beach, the section of Hutchinson Island just below the Fort Pierce Inlet. Some locals had rushed back across the Intracoastal to get guns but were too late to do battle. Bell, however, was reasonably certain that it was probably not the same serpent spotted by her parents four decades earlier, in 1881.

## The Titusville Serpent

In December 1885, the first railroad arrived in Titusville, the Atlantic Coast, St. Johns & Indian River Railroad, which was then almost immediately leased by the Jacksonville, Tampa & Key West Railroad. The railroad now connected Jacksonville to Titusville. Passengers and goods traveled by steamer from Jacksonville along the St. Johns River to Enterprise, where they boarded the train to Titusville where the Jacksonville, Tampa & Key West Railroad had continued the line on to the Indian River Steamboat Pier, extending 1,500 feet into the Indian River where the river was deep enough for the steamboats to dock. There, the railroad-owned Indian River Steamboat Company continued the journey, picking up and discharging passengers and goods at stops along the way, traveling down the river to Jupiter, where trains continued south. Titusville essentially became the transportation hub of the Indian River.

The Indian River Steamboat Company's crown jewel was the *St. Lucie*, a 22-foot long sternwheeler. With 14 staterooms and ample dining and socializing space, she was considered opulent. More importantly, she only drew 35 inches of water, perfect for the naturally shallow Indian River. Titusville's role as a transportation center ended after only a decade, its demise caused by Henry Flagler's Florida East Coast Railway Company. Flagler's track reached Titusville in January of 1893, meaning there was now a direct route from Jacksonville that didn't involve ships in the St. Johns River. Flagler completed his track along the Indian River Lagoon, a total of 115 miles, by March 1894.

Flagler's vision of his railroad was a vertical monopoly—he would use his trains to bring his passengers to his hotels. So, when Rockledge hoteliers refused to sell their successful resorts to him, Flagler tore out the spur into town, destroying the tourism trade in Rockledge. Similarly, when Titusville landowners tried to gouge the price of land, Flagler chose to build elsewhere, buying up property in a sparsely populated stretch of coast known as Palm Beach.

By the end of 1894, tourism was bypassing the original resort towns, heading to Flagler's resort in Palm Beach. Out on Merritt Island, hotelier Peter J. Nevins was getting anxious about diminishing reservations. Nevin's Riverview should have had more reservations, having the advantage of being located nearly across the river from Cocoa Railway Station, the only depot left that Flagler permitted to offer transport to the Merritt Island resorts.

Nevin's Riverview was surviving, but he was concerned. And perhaps he had a reason—if the business failed, it was his name on the deeds even though Niven's Riverview belonged to his brother, Brooklyn Fire Chief Thomas F. Nevins. Chief Nevins, a Tammany Haller who retired in 1894 at age 50, opened a stock brokerage and built Riverview as his winter home, conveniently large enough to serve as a hotel or as a winter destination for dozens of Brooklyn politicians. Peter Nevins, however, took his worrying to the next level. He had sent letters to the editors of various newspapers complaining about the northern newspapers making up tales about Floridian fauna—giant snakes, killer insects, and mysterious animals. His concern was that such nonsense would discourage tourism. Considering his brother was still filling the hotel each winter with his entourage of politically connected Brooklynites, Peter Nevins sounds more like a crank than a concerned hotel proprietor.

In the April 26, 1895, issue of the Titusville newspaper, *Florida Star*, Nevins questioned the newspaper's editor as to whether a recent, rather sarcastically title article he had read in the New York papers was true. The article, Nevins complained, appeared to be a March 24 news dispatch from Titusville to the *St. Louis Globe-Democrat*. Nevins included a clipping from the April 7 *New York Sun*, which the *Florida Star* reprinted verbatim.

> Titusville, Fla., March 24. -- For a month or more there have been reports that there was an immense sea serpent in the Indian River, which showed a disposition to fight when molested, but these reports, up to yesterday, were regarded as the product of the overwrought imaginations of rivermen. Yesterday, however, the truth of these reports was confirmed by the appearance of the monster off this place.
>
> About 9 o'clock yesterday people on the wharf waiting for the steamer saw a great black object resembling a hogshead barrel floating in the river about seventy-five yards from shore. The object appeared to be lifeless and those who saw it thought it was a piece of wreckage. Capt. Simmonds and Fred White resolved to investigate. They took a boat and rowed toward the object. When within twenty-five feet of the object the men were surprised to see it show signs of life, and a moment later were horrified when a wicked-looking head, with basilisk eyes, was darted at them with a hiss that could be heard half a mile. The men backed water for life, and the monster began to uncoil itself and move. It went through the water like a snake, was about sixty feet in length, and its body in the thickest portion was as large as a barrel. The head of the monster was similar to that of a snake, and for about six feet along its back there appeared to be a row of fins. The body

of the reptile tapered gradually to a pointed tail. The monster moved down the river in plain sight of hundreds of people who were on the wharf. As it passed the men who had guns began shooting at it, and the reptile resented these shots by erecting its head six feet or more and emitting several hisses. Then it sunk below the surface and was seen no more.

Capt. Simmonds and Fred White, who went out to inspect the object, were so overcome when they reached the shore that restoratives had to be applied. They say that they saw rows of immense teeth in the reptile's mouth, and that its breath was most noxious. About midday a steamer arrived from the south and reported passing the monster thirty miles below Titusville. The appearance of the monster has demoralized tourist travel on the Indian River, and the houseboats of the wealthy northerners have been deserted.

The editor of the *Florida Star* amended a note after the article, saying it was "pure exaggeration emanating from the mind of some egotistical newspaper correspondent"—not exactly a denial, but perhaps the careful wording infers they suspected who the local source was. The two names mentioned, Simmonds and White were not randomly made up—both lived in Titusville.

Unbeknownst to Nevins, the article had already been picked up by other papers, indicating that the St. Louis paper had contributed the piece to the Associated Press telegraphic service. The way the *New York Sun* credited the story (Titusville, Fla., March 24.) would be correct if the story originated in Titusville, which it could not—the *Florida Star* was not a subscriber to the Associated Press. All other newspapers credit the tale as "Titusville corr. St Louis Globe-Dispatch," which is also incorrect.

The original *St. Louis Globe-Dispatch* article is essentially the same article as the version that ran in the *New York Sun*—with one notable difference. The story is sourced as a "Special Dispatch to the Globe-Democrat," a catch-all term. It could be someone in St. Louis who had recently returned from the Indian River area, or a letter home to St. Louis that was shared with the newspaper, or a crafty brother in Brooklyn drumming up a little free press by playing a practical joke on his brother on Merritt Island.

In retrospect, stories of oversized snakes were the least of Nevins's problems. By the end of the year, it was not the serpent in the river, but Flagler's railroad extending to Miami that had driven the Jacksonville, Tampa & Key

West Railroad and the Indian River Steamboat Company out of business. Peter Nevins advertised extensively in the New York papers, but it was his brother's political winter soirees that kept the hotel solvent.

Thomas Nevins also wasn't concerned about the hotel's success. He began buying orange groves in 1898, and Nevins Fruit Co., Inc. became one of the first shipping associations in the state. The company still thrives under the Parrish family, who bought out Thomas Nevins. The sea serpent hasn't been seen again since Peter Nevins wrote his terse letter to the editor.

**The Okeechobee Serpent**

The late 1890s saw a drought that all but wiped out agriculture in southern Florida. Even worse, the hunting and fishing packages touted by the hotels found the rivers too shallow to navigate with the tourist-friendly steamships. Excursions to Lake Okeechobee from Kissimmee were impossible. If both tourism and agriculture failed, Florida was doomed.

Others found a windfall in the face of the disaster, such as Buster Ferrel. Ferrel had a camp on Okeechobee Beach at the mouth of Taylors Creek on the north shore of Lake Okeechobee. His fortunes had improved by the Kissimmee River becoming navigable; he served as a fishing and hunting guide and sold wood to refuel the steam engines. Ferrel had been at his camp since the late 1870s when the north shore of Lake Okeechobee was beyond the frontier, deep in inhospitable sawgrass, impassable swamp, and unfriendly Seminóles. And as he had done before, Ferrel adapted.

And in the middle of Florida's drought, the most profitable business was alligator hunting. The drought had made hunting the reptiles easier, and alligator skin leather was a popular trend. Alligator hides had become such a significant source of revenue in the south that the U.S. Department of Commerce began tracking it as a separate industry—by 1893, Florida alone was shipping close to one million alligator hides per year to leather workers up north.

The prices had been dropping by 1901, even after the drought had subsided. Buster was netting $1 for a seven-foot hide and, if he was lucky, 10¢ for one under four feet. This was down by a third from prices the previous decade. The leather had fallen out of favor, so demand was dropping, which was fortuitous for the alligators. The species had been decimated to the point that Florida legislators were discussing how to protect the species from extinction.

But on Lake Okeechobee, it was business as usual for Ferrel. Fishing and hunting expeditions were starting to filter back down the Kissimmee on boats with a shallower draft. On one of his hunting surveys, he encountered a track pushed through the marsh. He assumed, based on the location, it was a massive alligator, bigger than the 12-to-14-foot adult males that used to be so plentiful.

His ego as a hunter mandated he track this behemoth (and potential revenue source). After several visits, it was apparent the beast frequented the path. Ferrel noticed a nearby cypress tree. For the next two days, he scaled the tree and used it as an impromptu tree stand. On the third day, he saw what made the trail—it was not a giant alligator. It was a giant serpent, one that Ferrel estimated was 25-to-30-feet long, 10-to-12 inches in diameter at the neck, and as large around as a barrel ten feet further down. The mouth surrounded by barbels like a catfish. The back was dark and the underside a dingy white. The creature frequently stopped and raised its head above the sawgrass, which typically grows to six or seven feet tall.

As stunned as he was by this unexpected appearance, he recalled that such a beast was a tradition among the Seminóles of the region. When Ferrel recounted the Native version, he claimed the animal was snakelike in appearance with "ears like a deer." This may be a mistake on Ferrel's part—snakes don't have external ears; alligators and crocodiles have external ear openings, but these opercular flaps are skin folds that seal to keep water out and would be barely noticeable above water. However, if Ferris meant "horns like a deer," he is describing the Great Horned Serpent that appears in the mythologies of many Native Americans, including the Seminóles (Chapter 15).

Ferrel's second reaction was to wait until the creature stopped in range and shoot it in the head. The serpent fled at a tremendous speed while Ferrel emptied his magazine at it as it disappeared into the swamps. Four days later, he ventured back to see if the track had been used. Ferrel noticed a flock of buzzards in the distance. He had killed the giant snake. The buzzards had fed on the carcass to such an extent as to make the skin irretrievable. But he took the head, which he hung in his home on the Kissimmee. It was described as 10 inches from jaw to jaw and filled with razor-like teeth.

*The New York Times*, not a local paper, reported the story. This suggests Ferrel told the original story to one or more of his hunting/fishing clients from up north, who returned to New York with the story. The story closed with the claim that Ferrel had headed back to find the carcass and save the skeleton for the Smithsonian. The Smithsonian is still waiting.

**The Zimmerman Case**

The Sunday newspapers in August 1934 were having a field day with a syndicated article about a case in the Fifth Circuit Court of Appeals. Back in September of 1929, Nellie F. Zimmerman brought two separate suits on denied claims for the life insurance policies on the life of her late husband, Albert H. Zimmerman.

Albert Zimmerman, a Clearwater businessman, disappeared in the Indian River on August 24, 1927. He had been taking sand samples along the shore, looking for a specific type of sand that would determine where a glass factory could be built. His car and street clothes were found along the bank, and his sample scoop was planted in the sand 300 feet out in the shallows. Based on the location, his family believed he had drowned and his body swept out to sea through the nearby San Sebastian Inlet. The story was covered extensively by the *Tampa Tribune,* whose readership included Clearwater.

The insurance claims had dragged on since 1929. Albert Zimmerman had four life insurance policies with different companies, so the potential payout was significant—$165,000. The bulk of that amount was through Mutual Life Insurance—a $25,000 plan with a double indemnity for accidental death. Nellie Zimmerman wanted interest and civil penalties as well, putting Mutual Life on the line for $85,000.

Since Zimmerman's body had never been recovered, all four insurance companies were disinclined to pay. Mrs. Zimmerman gave the insurance companies two years to prove he was still alive or recover his remains. When the companies still balked at paying the policies, Zimmerman went to court. Mutual Life, in particular, mounted an aggressive defense, but seven years, four trials, a hung jury, a retrial, and two appeal courts later, the widow had won all four suits.

The press coverage of the February 1933 trial noted that Mutual Life had probably spent more money on searching for Zimmerman and fighting the case than they would have had they simply given Mrs. Zimmerman a check in the first place. Albert Zimmerman's financial history was exposed as a man teetering on the edge of financial ruin, so Mutual Life had tried to convince the jury that Zimmerman was a suicide, not an accidental death. Zimmerman's team countered with the testimony of local fisherman Charles Smith. Smith had been part of the search parties scouring the river and noted he had seen eight-foot sharks and three-foot stingrays near the inlet. He further testified he had heard of shark attacks in the area and noted a retired judge who, after being struck by a stingray barb, was in constant pain for the rest of his life. The case ended in a hung jury.

The 1934 retrial got ugly from the start. Mutual Life brought in eyewitnesses who claimed to have seen Mr. Zimmerman after his alleged death, one spotting him in Belize and one in South Carolina. Both witnesses were easily proven to be at best unreliable, casting the insurance company's trustworthiness into question. The Zimmerman lawyers countered with several fishermen from the Sebastian area who recalled finding a human femur in the river the year after Zimmerman disappeared. It was green with age, covered in barnacles, and found three miles away, but it was allowed into evidence.

The insurance company maintained that Zimmerman was unhappy in his marriage, about to be exposed for financial improprieties, and had faked his death. Mrs. Zimmerman maintained he drowned and that sharks and other sea creatures ate his remains, or that his body was swept out to sea. The phrasing of Mrs. Zimmerman's position, as quoted in the *Tampa Tribune,* would prove to have ramifications.

The retrial was ugly but brief. Two days later, Zimmerman was awarded her husband's life insurance of $25,000 from Mutual Life and $20,000 from New York Life Insurance. The insurance company was not required to pay the double indemnity—accidental death had not been proven.

The press following the trial had been sympathetic to Mrs. Zimmerman's side, thanks to the Tampa newspaper, which had been providing updates via the Associated Press. But the *Tampa Tribune* quote attributed to her about her husband's body having been "eaten by sharks and other sea creatures" was irresistible. The trial ended June 2, 1934, amid a spike of sightings of the Loch Ness monster and weeks after the iconic "surgeon's photo" of the Loch Ness monster had been revealed.

King Features Syndicate created a three-quarter-page spread of photographs and drawings, emphasizing the salacious details such as Zimmerman's infidelities and impending financial ruin before bringing up the sea serpent, including pictures of a recent carcass washed ashore in Cherbourg, France, and a skate, modified and dried, known as a Jenny Haniver. What had been a long trial about insurance was suddenly a tabloid story about a sea serpent.

Mutual Life and New York Life Insurance Company appealed to the US Fifth District Court of Appeals, and this time the court overturned the award because of how extensively dependent upon circumstantial evidence the court decision was based upon, and how poorly the court had dealt with such evidence.

That case, "Mutual Life Ins. Co. of New York v. Zimmerman" remains a cited case for civil suits relying on circumstantial evidence. Mrs. Zimmerman ended up with less than $60,000, from the two smaller policies that chose not to pursue further appeals. As for Albert Zimmerman, whether he fled his old life, drowned, or was a sea serpent's meal remains a mystery.

# CHAPTER 9
## Lake Monsters

"Lake Monster" is a term similar to "Sea Monster" in Floridian parlance—a catchall term of hyperbole. Lake monsters have their own quirk—they tend to be anything but cryptid in nature. When a seven-foot alligator was pulled out of Goodthing Lake (now Lake Earl) in Fort Walton Beach back in 1966, the local newspaper disappointedly headlined the story "Goodthing Lake 'Monster' Really is only Alligator." The inference is that given a few more years, the lake monster's tale would have evolved beyond a simple alligator.

**The Orange Lake Monster**
The September 1903 issues of Ocala newspapers included an article from the Island Grove correspondent of the *Gainesville Sun*. E. P. Cail was planting trees on an Island Grove citrus farm along Orange Lake when he heard odd noises in the water. Going to the shore to investigate, he discovered waves in the water, more than two feet high.

Suddenly, something leaped out of the water, dove back into the water, and disappeared. Cail described it being 15-to-20 feet long, with red eyes, ears like an elephant, a fish-like tail, and a head like a dog. Cail was convinced it came into the lake via an underground current from the ocean.

Orange Lake, 20 miles southeast of Gainesville, covers 12,550 acres, the largest lake in North Central Florida. However, Orange Lake only averages 5.5 feet deep with a maximum depth of 12 feet—not optimal for the traditional image of a sea monster.

Cail's conviction that a subterranean tunnel from the lake led to the ocean is a widespread misconception people hold about many lakes. The belief in underwater passageways is a widespread occurrence across the globe in legends of lakes monsters. Folklorist Michel Meurger suggests such secret tunnels to the sea date back to the idea that monsters were guardians of entrances to the underworld. Orange Lake has several sinkholes in the southeast corner of the lake that could pass as cave openings. At that time, the water in Orange Lake flowed into Orange Creek and then into the Oklawaha River which emptied into the St. Johns River. Outflow today is controlled by a fixed-crest weir located in the southeast portion of the lake that regulates water flow into a reservoir.

E. P. Cail was a nurseryman in La Crosse, north of Gainesville, and 40 miles from Island Grove. La Crosse was an agricultural community with only small ponds. So Cail was unfamiliar with lake denizens that could leap from the water. Leaping out of the water is a trait of the Atlantic sturgeon (*Acipenser oxyrinchus oxyrinchus*).

Atlantic sturgeon live in rivers and coastal waters along most of the coast, from Canada to Florida. They hatch in freshwater rivers and head out to sea, only returning to freshwater to spawn—and Orange Lake had ocean access. With bony plates known as scutes along the body and barbels on the snout, the living fossil could easily become the startled nurseryman's lake monster. A dog-like head becomes the pointed sturgeon's head with the barbels mis-identified as a dog's whiskers and the pectoral fins being mistaken for hanging elephant ears. Atlantic sturgeon have been recorded up to 18 feet in length, and for reasons ichthyologists don't fully understand, they can leap out of the water up to seven-foot heights.

The largest sturgeon ever recorded measured 24-feet long and weighed 3,463 lbs. While that certainly qualifies as a lake monster per se, it is not a lake monster of the cryptid variety.

## Lake Clinch

Cryptozoological publications that mention the Lake Clinch serpent invari-ably cite *A History of Polk County* (1928) by M. F. Hetherington. The entire reference is a paragraph on page 154, part of a history of the town of Frost-proof, which sits between Lake Clinch and Reedy Lake.

> There is a tradition that a sea serpent, or a lake serpent, used to haunt Lake Clinch. The Indians many years ago insisted there was an immense serpent in this lake. In 1907 residents of Frost-proof declared they had seen the monster, and that it must be thirty feet long—this, too, before post-prohibition liquor was known.

Hetherington, the successful publisher of the *Lakeland Evening Telegram*, had only lived in Lakeland for 16 years when he started his county history as a retirement project. As a comparatively new transplant, he relied heavily on a board of advisors with generations in the county. More telling, his introductory notes include the fact that he went to all of the county newspapers and read their archives, some already brittle with age, and that he mercilessly edited material because of size constraints. One of the items trimmed out was the source of his lake monster report, a July 25, 1907, edition of a rival newspaper, the *Bartow Courier Informant*.

In this earlier piece, many more details of the lake monster come to light. An unsigned correspondent reported he was crossing Lake Clinch with a hired hand to go deer hunting. Just as he cleared the shore, the hired hand exclaimed: "Great God, what a turtle's head!" The creature then disappeared before the anonymous writer could secure his oars and turn to look. He had heard there was a "sea serpent" in the lake and was disappointed to have come close to seeing it. They waited around, hoping for a return visit, in vain, before continuing across the lake.

The correspondent expressed doubts as to the creature's existence, since none of the various citizens who had reported having seen the serpent could fully describe it. All the sightings were partial, such as his hired man only seeing the head rise out of the water. But he had to admit that had changed in the last few months because the most recent witnesses were of such an unassailable reputation that the beast had to be considered real. These witnesses were John Schoonmaker, Frank Jacques, and Harley Brackin. They claimed they had all three seen the creature swimming on top of the water. Jacques and Brackin were reluctant to discuss the matter. They were successful citrus farmers with extensive holdings and preferred keeping their reputations intact.

Schoonmaker was the youngest of the three and merely the son-in-law of a third citrus farmer. And even he was uneasy repeating his tale, fearing he'd be accused of having an overactive imagination. Schoonmaker said they saw the creature as a storm was rolling in. A bolt of lightning lit the sky, and the ensuing thunder startled the serpent, which lurched in a sudden turn in the opposite direction, exposing a 30-foot torso as it fled.

In a thunderstorm, it could be challenging to tell whether the fleeing beast was really nothing more than rolling waves on the dark surface, but it is important to note that none of the three witnesses were Florida natives. Schoonmaker was from New Jersey, Jacques from Connecticut, and Brackin from Alabama. So, even though it is unlikely in a town that sits between two lakes, an alligator might not be recognizable at night in a choppy lake. But the answer may lie in a neighboring lake.

In 1891, an outlet ditch was dug connecting Lake Clinch and neighboring Crooked Lake to lower the shoreline of Crooked Lake, producing muck lands for the Florida Sugar and Rice Company. There is a fall of six-to-eight feet between the mouths of Crooked and Clinch Lakes, and local property owners protested the drawing of so much water from the lake. The neighbors began blocking the water, and a temporary injunction was obtained to keep the ditch closed. The ditch's opponents had the time to buy the land along the canal and built a dam. The dam was dynamited, rebuilt, dynamited again. The ditch was finally opened in 1922.

Within months of the two lakes being connected, Crooked Lake had its own lake monster sighting. The *Tampa Tribune* proclaimed a "prehistoric monster" was allegedly spotted in Crooked Lake. Reporter Worthington noted multiple sightings and included the description of an eyewitness. It was initially assumed to be a large alligator, a well-known resident of the lake, but as the creature sped toward the boat, it became obvious that it was not an alligator. It was described as a 20-foot long creature with a three-foot-high fin on its back that swam fast enough to throw a wake "like a powerboat." The snout was narrower than an alligator's, and it swam with agility and without a serpentine motion. It came within 100 feet of the boat, then veered away and disappeared.

The *Tribune* was partly correct—it was prehistoric. The Crooked Lake description matches that of a living fossil—the Alligator Gar (*Atractosteus spatula*), one of the largest freshwater fish in North America. The alligator gar can weigh up to 350 pounds and reach 10 feet in length. The description of a fin on its back would then refer to the dorsal fin positioned toward the back of the body (rather than the more common middle of the back), leaving the justifiably panicked eyewitness to extrapolate a longer length.

Up until this point, the Lake Clinch Monster was a local legend. Then a 1939 article by local historian Sophronia Carson Ohlinger added additional eyewitnesses, including herself. Her brother-in-law encountered the creature while in the lake, noting it was snake-like in appearance except for the head, which was not only as big as a human head, additionally also had a similar shape, i.e., more rounded than snake-like. This echoes the original 1907 article describing a turtle-like head. She also talked to Frank Jacques from the 1907 sighting, and his story had not changed over the decades. Perhaps the most interesting observation made by both Jacques and Ohlinger is that the sightings seemed to occur when a storm was rolling in.

An approaching storm would lower the barometric pressure, which in turn decreases dissolved oxygen levels near the surface. With activity and distribution patterns affected, the situation would present feeding opportunities for piscivores such as alligator gars.

Regardless of its identity, the Lake Clinch Monster languished as a local legend, in spite of attempts to raise interest in the 1970s through the 2000s with occasional and increasingly tongue-in-cheek newspaper articles, including an ill-conceived effort to rename the cryptid "Clinchy" to draw associations to Nessie. That changed in 2011 when cryptozoologist Scott C. Marlowe published a book claiming that the Lake Clinch Monster was responsible for the 1926 drowning of C.M. Mallett.

Charles Mercer Mallett was a successful businessman and a highly regarded civic leader in Frostproof. Mallett had announced on August 30 that he was heading off on a business trip in the morning, so his absence was not noticed until September 2, when a local fisherman snagged his body on his hook and pulled him of out of Lake Clinch. The newspapers politely declined to offer gory specifics other than that the identification of the remains needed to be made through dental records and his ring. Mallett had last been seen Monday evening, indicating he had been in the water for less than three days. His remains had been pulled out the water near the boat launch where he would have boarded his small boat for some last-minute fishing. The water temperature would have been in the mid-70°F range, slightly cooler than usual due to heavy seasonal rain.

From a forensic perspective, Mallett had not been in the water long enough for the type of decomposition that would make him unrecognizable. And even in the warm lake, putrefaction would not have progressed to make his remains sufficiently buoyant to float. So, his body was probably snagged resting on the shallow bottom—Lake Clinch is bowl-shaped and gradually deepens. It would require a powerful cast by an angler to reach even 10-foot-deep water from the shore.

The newspapers in Orlando and Tampa covered the drowning extensively, and specific details were shared about the well-known businessman. He went out in the evening to fish, accompanied only by his dog. He was a poor swimmer but was wearing a swimsuit, which may suggest he had fallen out of his small motorboat more than once in the past. In 1926, a man's swimsuit was a woolen tank top over a snug-fitting pair of shorts just above the knee. More importantly, it could weigh as much as 10 pounds when wet.

By falling off a boat in the dark, with Florida's fall weather pattern of sudden localized downpours, combined with the extra weight of the wet swimsuit, Mallet faced a combination of factors that would not benefit such a poor swimmer's survival odds. A lake monster is not required in this tragedy.

The body sank to the bottom where local scavengers quickly found it—catfish, snapping turtles, and crawdads would all be attracted to the soft tissue of the face. If Mallet's death wasn't due to the Lake Clinch Monster, Marlowe suggests, alternatively, that it was due to a South American anaconda, specifically a green anaconda (*Eunectes murinus*), the only species that reaches the 20-to-30-foot range. This identification is circular reasoning. Marlowe believes the presence of anacondas in 1926 proves an archaeology site in nearby Lake Pierce is the remains of a "Mayan trading post." Since the Mayans worshiped anacondas, they brought some with them for religious and security reasons.

The site in question is 10-acre Snodgrass Island in Lake Pierce. And Marlowe's sources were severely out of date by the publication of his 2011 book. As early as 1993, archaeologists had noted Snodgrass Island was home to a prehistoric burial mound from the Late Archaic Period (3000-500 BC) and a Seminóle ceremonial site. The site appeared regularly in the media from 1994 when the owner struggled with the state to assume ownership at a fair price, rather than see the island developed. The State acquired Snodgrass Island in 2009 and integrated its oversight into the Allen David Broussard Catfish Creek State Preserve. None of the assessments or surveys mention Mayan ruins or anacondas in the 8,000+ acres of the preserve.

The Lake Clinch monster is occasionally reported, but the sightings are now rare. The lake was named after Fort Clinch, established on the northern shore of the lake in 1850, which in turn had been named for the recently deceased General Duncan L. Clinch, a military mastermind who was so unqualified to command that he is considered the primary cause of the First Seminóle War. The name change was not popular, and it would take another 50 years before maps stopped referring to it by its original name, Lake Lochapopka. Lochapopka is from the Muskogee, roughly meaning "where the turtles eat" or the "turtle eating place." Based on the lake's history, it might be helpful if the name had specified who or what was eating who.

## Lake Tarpon

The US Army Corps of Engineers began work in 1965 on a channel from the south end of Lake Tarpon to Old Tampa Bay. When the lake is at a safe level, the flood gates are left closed to prevent saltwater from entering the freshwater lake. When the water reached a certain level, the flood gates are opened and water flows down the canal and into the bay to prevent flooding.

As it turns out, when the gates are opened to let the water out, it also allows in visitors. Manatees have repeatedly wandered into the lake, much to the concern of wildlife officials, not only because of boat traffic but because Lake Tarpon's winter temperatures are fatally cold to manatees. The sirenians must be either coaxed back out the gates or captured and relocated. When one arrived in 2003, Florida Fish and Wildlife Conservation Commission told the *Tampa Bay Times* that their biggest concern was that boaters, unfamiliar with manatees in the lake, might injure the creatures.

This concern over unfamiliarity was prescient. Less than three years later, an entrepreneur launched a website devoted to "Tarpie," the Lake Tarpon Monster. The website was firmly tongue in cheek, with such bon mots in the disclaimer as "The information presented is believed to be correct and accurate, but the author also believes in Santa and sea monsters."

The website painted a variety of possibilities ranging from a giant catfish to an ancient alligator but stressed that, whatever it was, it was a man-eater. The website's position was that Tarpie was a descendant of a long-necked plesiosaur with a decidedly alligator-like head, but now was an ornithopod, an upright, bipedal dinosaur (possibly meant as a reference to Ivan Sanderson's initial belief that the Clearwater Three-Toes was a dinosaur; see Chapter 13). Tarpie was also three-toed, inferring that Three-Toe had relocated. The rest of the page was a link to an online gift shop of tote bags and tee shirts.

In spite of the flippancy of the site, Lake Tarpon almost immediately began appearing on lists of the world's lake monsters, with several cryptozoological news aggregator websites adding the bipedal reptile reports as well.

An even rarer visitor was removed from Lake Tarpon in 2013—an 11-foot, 700-pound American crocodile (*Crocodylus acutus*) that was far north of its normal range. The timing was unfortunate, as the Lake Tarpon monster website had shut down the year before.

**The Lake June Monster**
In 1965, Florida's Lake Placid was still relatively undeveloped and remote, home to barely a thousand year-round residents. It was mostly known as a crossroads for fishing enthusiasts. Earlier in the year, Lamonte Moore, the editor of the *Lake Placid Journal*, was approached by an out-of-state fisherman who claimed, while angling on Lake June-in-Winter (the largest of the lakes that ring the town), that he'd tangled with a giant catfish, estimated at 12-feet long with a two-foot-wide head.

The *Journal* dutifully reported the claim, which was picked up by the news wires. The residents with homes along the shore expressed doubts, having spent their lives fishing on the lake without ever having encountered a 600-pound monster. The Lake Placid Lions Club saw an opportunity to fundraise for their charities and had a "Monster Marathon" fishing contest over the Independence Day holiday, drawing hundreds of fishermen trying to catch the behemoth at Lake June Park.

By the second year, the Lions Club festivities had expanded to include a barbecue (and fish fry, of course). But when the *Tampa Tribune* began expressing doubts, suggesting the Lions Club had invented the Lake June Monster for publicity purposes, the uproar forced a second article the next month. This one admitted there had been reports from eyewitnesses, but still wouldn't commit to the possibility of such a monster in the lake. In any case, it didn't slow down the annual fishing contest, which added food vendors and children's activities.

In fairness to the *Tampa Tribune,* a catfish of that magnitude was suspicious. Lake June-in-Winter is one of the deeper water lakes in the region, with depths down to 40 feet so that a large catfish could remain hidden in the lake. But the issue is how big—in 2019, the record for catching the largest catfish in Florida history was a 47-inch-long, flathead catfish weighing 69.3 pounds. A nine-foot catfish (646 pounds) was caught in Thailand, but that fish belongs to the species known as the Mekong giant catfish, which has not been identified as a Florida species, invasive or otherwise.

In 1977, the *Tampa Tribune* revisited the Lake June Monster. The Lions Club had moved their barbecue into town as the population and demographics changed. The paper mentioned several new accounts, including one from a determined sportsman who rented a helicopter and flew over the lake with a baited grappling hook in an unsuccessful attempt to catch a giant catfish with a giant fly-casting rig.

Also of note is that the creature was now called "Old Moe." This name was a new facet of the story. All the original press referred to the fish as the "Lake June Monster." The locals who spoke to the *Tribune* reporter were blurring two memories into one. "Old Moe" was the name of a giant barracuda in a relatively obscure move called *Raymie* (1960), where Raymie (David Ladd), a nine-year-old boy and avid fisherman, tries to catch a legendary giant barracuda known as Old Moe. When he finally manages to hook Old Moe, he has a change of heart, thinking that some legends are best kept alive. The name Old Moe is based on regional fishing slang. In California, where the movie was filmed, a "moe" was the one that got away.

*Raymie* was popular in central Florida, where fishing is a lifestyle. In the years before the Lake June Monster surfaced, the film ran regularly. After its first theatrical run, it simultaneously ran at drive-ins, kiddie matinees, and on television. It was such a fixture in the area during the early 1960s that it's quite reasonable that Old Moe would be associated with the name of the legendary local giant fish.

The last sighting of the Lake June Monster was off the lake's northeastern shore at Breezy Point in 1976. This is near the border of Lake June-in-Winter Scrub Preserve State Park, an 845-acre preserve to protect sand scrub, one of the state's most endangered natural ecosystems. Perhaps it also protects a local legend.

## The Lake Zephyr Monster

In January 2003, Gary Hatrick, the associate editor of the *Zephyrhills News,* was on the shore of Lake Zephyr, interviewing Steve Seneels, the founder of the annual Celtic Festival. This was the third year of the festival and the list of events was unremarkable. Hatrick knew he needed something to spike interest in the festival. With the flexibility afforded by a locally owned newspaper, he had an idea. He had Sennels meet him at Lake Zephyr for a photo to accompany the article. He had the unsuspecting Seneels point out toward the lake. Hatrick took the shot and then went looking for wading birds. He found a species he didn't recognize and took a few shots of a bird swimming in the shallow water. Back at the newspaper offices, he had the bird digitally doctored to be blurry, then placed it, in the style of the Loch Ness "surgeon's photo," just beyond the pointing hand of Seneels. The article may have been a listing of events, but the accompanying photo showed the Celtic Festival's founder pointing toward a blurry image reminiscent of a sea serpent. And thus was born the legend of Zeffie, the Lake Zephyr monster.

The town loved the gag (although not the Zeffie tee-shirt, which didn't sell). And Hatrick told the truth when he said he didn't know what the creature in the photo was. When the 2004 festival neared, Mayor McDuffie joined the fun, declaring the body of water in Zephyr Park would be called Loch Zephyr for the festival. With his tongue planted even more firmly in his cheek, Hatrick created a backstory for Zeffie, now the mate of Nessie, separated when Pangaea split apart millions of years before. By 2005, the lake monster was a part of the festivities. There was a children's "Draw Zeffie" contest, and an inflatable Zeffie was placed in the lake to see if the real one could be lured out to make an appearance.

Unfortunately, the lake monster mascot became difficult to integrate into the festival when it moved away from the lake to neighboring Dade City in 2011. The move didn't agree with the locals either—2012 was the last Celtic festival.

No one has sighted Zeffie since. The current (but perennially tongue-in-cheek) theory is that Zeffie, which had utilized a connecting underground waterway to Crystal Springs to avoid the times Lake Zephyr went dry in the summer droughts, has relocated permanently to Crystal Springs. Hatrick proposes that Zeffie was the real reason behind the closing of Crystal Springs Recreational Park and the educational center now there for students is just a front for Zeffie research.

It took less than a decade for the least discerning cryptozoologists to start including Zeffie with cryptids such as Nessie, Champ, and Caddy. Apparently, cute nicknames are more validating than actual research. Hatrick notes there are discussions to bring back the Zephyrhills Celtic Festival. Whether it returns to Lake Zephyr will determine whether Zeffie will also return from retirement. There's no word on the tee-shirts.

### Lake Norman

Lake Norman is a man-made lake located north of Charlotte in North Carolina. Duke Power created it in 1959-1964 by the construction of the Cowans Ford Dam. It is used to power the hydroelectric station at Cowans Ford Dam, supplies water to the coal-fired Marshall Steam Station and McGuire Nuclear Station. With over 500 miles of coastline, it is a haven for swimming, fishing, camping, diving, and, of course, lake monsters.

In 2002, the website lakenormanmonster.com was launched to monetize "Normie," the Lake Norman monster. Complete with witness reports, links to Normie's brief television career, and Normie merchandise, the website essentially maintains an image of the creature as perceived by Matthew Myers, the website's creator. Myers has also taken the first ten years of sightings from his site (2002-2012) and put them into book form.

A cursory glance at the posted sightings show the majority are mysterious wakes on the water and obvious misidentification of large fish species know to inhabit the lake, notably alligator gars and catfish. And the fish are monsters (in fishing lingo). In 2004 angler Joel Lineberger caught a state record size blue catfish (*Ictalurus furcatus*), 85 pounds in weight, and measuring 51 ¼ inches in length and 35 ½ inches in girth. Although the state record has since been broken, it remains the largest catfish pulled out of Lake Norman—although several local old-timers believe there are 100+ pound catfish still in the lake.

Just by witness reports, it appears that the lake monster is actually a variety of very large fish, unfamiliar to casual recreational boaters and swimmers, both in size and species. Some of the more avid amateur monster hunters theorize that the Lake Norman Monster might be the result of a mutation caused by the nearby nuclear power plant, but it seems more likely that in a manmade lake, the lack of predators plus ample food allowed fish to achieve sizes that are rare in natural lakes.

### The Lake Worth Muck Monster(s)

The "Lagoon Keepers" are a Florida nonprofit group that works with residents and law enforcement to remove hazards from the Intracoastal Waterway and help remove litter from the water. In August 2009, members Greg Reynolds and Dan Serrano were out in a boat off Peanut Island, near Channel Marker 10, investigating a report of debris in the waterway.

They never found the debris because they saw something else—an unusually long rippling trail, the wake of a shadowy 10-foot-long creature just below the surface. When the two got within 10 feet, it dove deep and disappeared. The two were able to shoot video of the creature's wake. Because the creature had disappeared in the muck that passes as the sea bottom in the Intracoastal, they named it the "Muck Monster."

Reynolds was a member of the County Artificial Reef and Enhancement Committee and happened to discuss the odd wake at a meeting. Another member mentioned it to a friend who was a reporter at WPTV, a local television station. The station ran a story about the encounter, taking the footage to Florida Fish and Wildlife Conservation Commission Marine Biologist Thomas Reinert who could not identify the creature only by its wake, but noted that since the creature never broke the surface of the water, it eliminated dolphins, sharks, sea turtles, and manatees.

WPTV posted the video online and it went viral, drawing the attention of national newscasts and syndicated paranormal radio programs. It culminated in the "Top Ten Questions Received by Palm Beach County Authorities about the 'Muck Monster'" list on *Late Night with David Letterman*. The Lagoon Keepers, sensing an opportunity to convert the publicity into additional funding, began selling Muck Monster tee shirts. The coverage created a literal feeding frenzy in nearby waterfront areas. It was possible to stroll from restaurant to restaurant, dining on comestibles as Muck Burgers (hamburgers stuffed with guacamole and pico de gallo), Muck Monster martinis, and Muck Monster pizza.

The widespread embrace of the Muck Monster had to cause chagrin in one quarter. Less than four months before Greg Reynolds and Dan Serrano first spotted the Muck Monster, the History Channel had aired an episode of their *Monsterquest* series on a different monster in the Lake Worth waterway, one that had generated little to no local interest. It took until mid-August for the *Palm Beach Post* to write an article. Any potential interest would be quashed two weeks later by the Muck Monster.

The earlier account started in 1999 when Gene Sowerwine, a local fisherman, was casting near Singer Island in the Lake Worth Lagoon and saw something odd in the water. Since it was his preferred secluded fishing spot, he saw it several more times. He decided to set up a video camera, occasionally bringing friends and family to witness his "sea monster." Over the next decade, he taped hours of the creature surfacing and submerging. It was big, agile, and flippered, with smooth skin with mottled coloring. But most unusual was its three-pronged tail like a trident, which Sowerwine could not match with any known animal. After shopping around his footage in cryptozoological circles, the trident-tailed "Sowerwine Creature" was brought to the attention of the producers at *MonsterQuest*.

The producers interviewed cryptozoologists and scientists. One of the first interviews was with Dr. Martine de Wit, who unequivocally declared it was a manatee (*Trichechus manatus*) with a mangled fluke. Dr. Peter Sorensen of the Minnesota Department of Fisheries, Wildlife, and Conservation Biology leaned toward de Witt's sirenian identification. Others disagreed. Dr. Edward Petuch, a geologist at Florida Atlantic University, suggested a disoriented cold-water seal, such as a female Hooded Seal (*Cystophora cristata*). Cryptozoologist Scott Marlowe echoed a similar pinniped identification, suggesting it could be a surviving Caribbean Monk Seal (*Neomonachus tropicalis*), officially declared extinct in 2008. Both would require a similar theory of a hind flipper so badly mangled to appear three-lobed that it prevented the migratory seal from leaving the area for the decade Sowerwine had observed it.

Considering that Dr. de Wit ran the Marine Mammal Pathobiology Laboratory for the Florida Fish and Wildlife Conservation Commission and was the State's expert who works exclusively on Manatee rescues and necropsies, it was probably unnecessary for the program to hire a dive team in chainmail shark suits with military-grade sonar to explore the waters of the lagoon, where visibility is less than a foot on a good day.

The *MonsterQuest* program ended with a "forensic animation" of the creature created by combining multiple images from various images in the Sowerwine footage. The result was so similar to a manatee that Petuch also agreed it was a manatee, but he tried to save face by claiming the non-sirenian behaviors must be overlapping sightings from a coexisting seal community.

Any lingering interest in the Sowerwine Creature as a cryptid ended with a cold snap in late December. When cold temperatures hit, manatees seek out warm water. The discharge canals of power plants are particularly popular, and Florida Fish and Wildlife Conservation Commission uses the opportunity to count the manatee populations while they are congregated. Among the aggregation at Florida Power & Light Company's Riviera Beach power plant, researchers discovered one with a three-lobed fluke. The Commission concluded a boat propeller had struck the manatee's tail at some point, and the sirenian's fluke grew back into three separate prongs. This was the Sowerwine mystery cryptid, as de Witt had predicted. The press release noted the origin of the other Lake Worth cryptid remained unknown.

That other creature, the Muck Monster, continued to grow in popularity. The Palm Beach County Convention and Visitors Bureau had been posting Muck Monster articles on their website. When the hits on those articles broke four million, the County Commissioners began paying attention. By October, with the winter tourism season starting, the county elevated the Muck Monster into a marketing icon, making him an official resident, highlighting him at pier events, and installing coin-operated viewing stations. A local resident began making appearances as the muck monster, assuming the monster was actually a Spanish moss-covered, kid-friendly humanoid.

By 2011 interest tapered off when there were no further sightings. By 2015, the muck monster suit was relegated to occasional appearances promoting water safety before the costume was finally retired. In all the hoopla, no viable identification was ever offered as to what the Lagoon Keepers saw in 2009. However, unlike its similarly commercialized cousin in Lake Zephyr, the tee shirt is still available.

# CHAPTER 10
## River Monsters

River monsters are rare. If the river is shallow, visibility prevents anything hiding in it. Deepwater rivers are more conducive for cryptids, but increases in motorized boat noises are enough to drive any critter off in search of quieter territory. The rare river monster sighting is often described as an oversized snake, even larger than the swamp serpents of Florida's Treasure Coast (Chapter 8). So, either there were giant snakes in Florida long before Hurricane Andrew resulted in the current Burmese python infestation in the Everglades or a lot more traveling circuses misplaced their herpetological exhibits than expected.

### The Savannah River Monster

At 5 p.m. on July 31, 1820, a slave was chopping wood along the Savannah River just below Augusta, Georgia, when he encountered a "sea monster." The local newspaper described it as 60 feet long with a head that was six feet (!) in circumference. It surfaced, moving slowly but steadily, then submerged for about 10 minutes before vanishing. The report stated the creature was various colors, and the protuberances on its back appeared to be about six feet apart and about the size of a standard rum hogshead (48 inches long x 30 inches wide).

The keys to understanding this sighting are in the description using the "rum hogshead" as a comparative size and the opening statement in the article: "This monster of the deep has at last made his appearance in Savannah River." The report is satire, mocking the seemingly endless accounts in southern papers about the sea serpent off the coast of Cape Ann and Gloucester Harbor in Massachusetts, where the years 1817-1819 had generated hundreds of sightings.

The *Chronicle's* follow-up article in the next issue should have resolved the matter for even the most gullible. In this article, parody is replaced by farce. The sea serpent, attempting to get past the Sand Bar Ferry, got stuck and was trapped on the sand bar. The readers of that time would know this was nonsense—the river and the submerged sandbar was so shallow that the ferry was a flat-bottomed boat that was poled across the river.

A large group came to ogle the creature, including a group of local "Sons of Neptune," the nautical peers of the Revolutionary War's Sons of Liberty. Together with some backwoods hunters, the old salts approached the beast. With one stroke of his tail (described as a mile in circumference), the serpent "killed, mangled, or prostrated everyone within his sweep."

The carnage prompted a second party to attack. "[A]fter considerable skirmaging, an Alabamian, of herculean stature" straddled the monster and blinded it. The snake was heard to surrender with "Enough! fair play boys," helped down from the shoal with an Archimedes lever, and rolled into the city for those who hadn't seen him yet, "a trophy to Alabama courage and New England veracity."

*Augusta Chronicle* Metro Editor Bill Kirby resurrected the tale for his newspaper column in 2010, revisiting the story in 2014 both for his column and a video blog version. He treated the story as a news story from the city's past, neglecting to put the story in any context that would explain it was satire, such as the serpent talking. He also whitewashed the tale, first referring to the slave who allegedly first sighted the creature as "a man chopping wood" in the first article and as an anonymous Augusta laborer in the second. So, due to a lack of context, we have a fascinating cryptid encounter instead of the parody as it was intended.

## The Needhelp Serpent

Chokoloskee is a small town on an island (now connected by a causeway to the mainland). At the edge of what is now the Ten Thousand Islands National Wildlife Refuge and Everglades National Park, to the fishermen, trappers, and farmers eking out a living along the coast in at the start of the 20th century, it was an outpost of civilization where mail could be collected and sent, supplies could be purchased, and goods could be loaded on boats for transport to Fort Myers.

On October 4, 1908, Col. James Demere and Ervin Lowe arrived in Chokoloskee from the distant farm village of Needhelp on the Turner River. Named by the first farmer, the town had soil so nutrient-poor that he decided he was going to "need help" to succeed. On modern maps, Needhelp was just north of Tamiami Trail (US Route 41), west of Collier County Road 839 inside Big Cypress National Preserve. The journey from Needhelp to Choko-loskee was an arduous 10-mile trip in a shallow draught boat, dodging low hanging branches through alligator and snake-infested waters more similar to a swamp than a river.

Demere and Lowe were heading downriver, just two miles from where the Turner River begins to begin to widen as it nears its mouth at Chokoloskee Bay, when they noticed an unusually large log in the water. As they neared "the log," they realized it was moving. It was a massive snake leaving the river. The tail was leaving the water on the south side of the river when it raised its head four-to-five feet off the ground, 20 feet away from the shore. They described the head "as large as a nail keg" (approximately 17 inches tall by 10 inches wide). The estimated length of the serpent was between 50 and 60 feet with a body the size of a flour barrel (approximately 28 inches tall by 20 wide). The two men wisely kept moving, and the last they saw of the snake, it was heading southwest into the swamp.

There were no further reports of the serpent, although the obligatory non-specific "old Indian tale" was mentioned in the story when it was included in the Chokoloskee weekly news round-up sent to *Fort Myers Press* by their local correspondent. The article also notes that Demere gave an update on the potato and malango crops, noting that both were still doing fine, and were still above water. This seems to indicate some localized flooding that may have driven the serpent to look for slightly less swampy territory. Which again, could mean that giant snakes were running undetected in the Everglades long before the current Burmese python problem.

## Charlotte County Sea Serpent

In 1951, treasure hunter Merrick C. Faulk and a friend had been exploring Shell Creek off the Peace River. Coming back downriver, they saw a log lying across the creek that had not been there when they had gone upriver earlier. As they got closer, they suddenly realized it was an enormous snake. They tried to back away, but the boat's wake seemed to disturb it. The snake went up and over a 15-foot-high riverbank and disappeared. Faulk would later describe it as big around as a human and 40-to-45 feet long. Not sure he believed it himself, he rarely mentioned it. Only a decade later did Faulk decide to approach the local newspaper. He did so because he realized he was the first person to see the Charlotte County Sea Serpent.

Faulk may have been the first to see the most recent sea serpent, but reports of a sea serpent in the area date back to 1903 when a local named Kreamer took a shot at the head of a "monstrous sea serpent" in the Orange River that had been scaring the tourists. The Orange River is a tributary of the

Caloosahatchee River, which connects to Charlotte Harbor, the epicenter of the next cluster of sightings.

Starting in November 1962, the newspapers in Fort Myers and Tampa were fanning a frenzy over sea monster reports from Charlotte County. Initial reports were from the Myakka River near Punta Gorda. Asked about the sea serpent, Dr. Eugenia Clark, the head of the Cape Haze Marine Laboratories, said she had already received several reports of a sea serpent in the Myakka River in previous weeks. The only thing she could confirm is that whatever it was, the stories all agreed it exceeded 20 feet in length. Her preliminary theory was that it was an anaconda that had escaped from a traveling zoo.

She also casually mentioned that the New York Zoological Society had a standing offer of $5,000 for anyone bringing them a 30-foot sea serpent. That got people's attention. By the weekend, so many groups of "adventurous wildlife enthusiasts" were out searching for the sea serpent that there were designated search coordinators. The story was beginning to obtain mythic proportions—sightings were reported in the both the upper and lower Myakka River and across Charlotte Harbor in the Peace River, in the Intracoastal along Don Pedro Island. The newspaper kept referring to it as a sea serpent in the headlines, while carefully reminding readers it was an anaconda in the body of the article. There were so many sightings over such a wide area in a small timeframe that some of the local snake experts suspected there was more than one anaconda being sighted.

When the newspapers began digging into their files, they discovered that two snakes had escaped from traveling circuses. A 27-foot anaconda had been lost by the Rogers Brothers Circus while traveling through Charlotte County in 1951. There were ample opportunities for a snake to escape—the circus had a winter camp nine miles south of Fort Myers starting in 1951. Then in 1959, the cage for a 16-foot anaconda was damaged during a brawl at a traveling carnival. By the time the Punta Gorda police restored order, the snake was gone. The records of Ross Allen's Reptile Institute in Silver Springs were searched and revealed five-to-six serpent escapes in the previous 15 years. Whether this identifies the sea serpent depends on whether it was indeed an escapee or merely an addition to a preexisting population, considering sightings predate all the escapes by half a century.

By mid-December, the area was having a severe cold snap, so much so that the survival of the anaconda(s) was in question. The concern was for naught— two weeks later, one of the last reports came from one of the first witnesses, who said he had seen "a couple of six-footers that looked just like the big one."

Although expeditions continued into 1963, the public had lost interest, and sightings tapered off accordingly. The last sighting was in May when a couple again spotted a pair of sea serpents in the Caloosahatchee River. Pauline and Cecil Rhoades were on the North Fort Myers side of the river across from the Florida Power and Light plant when they saw the serpents swimming downstream. The creatures suddenly reversed direction and swam upriver and out of view. Mrs. Rhoades told the newspaper that the serpents were 36 inches in diameter, each showing six feet above the water. She only saw the head of one, which she said was a large as a human's, only flat like a snake. The newspapers continued to call it a sea serpent until the end.

### The Altamaha Monster

Georgia's Altamaha River, one of the largest rivers in the state, empties into the Atlantic Ocean at Altamaha Sound above Brunswick, less than 50 miles from the Florida border. The river basin is massive, second only to the Mississippi River. It extends over 135 miles from its origin at the confluence of the Oconee and Ocmulgee Rivers through the undeveloped countryside with no dams and minimal crossings. The river becomes a broad estuary of islands, marshes, dikes, canals, ponds, and abandoned rice fields.

In other words, it is the perfect place for a sea monster. And it does indeed have a sea monster; it has had one since February 16, 1981. The Altamaha monster was the invention of Larry Gwin, a former reporter turned eel fisherman, and David Newton, a local architect who could build a tale as easily as a building. Gwin's keen sense of humor had been sorely disappointed when a reporter from Jacksonville had profiled him in an article on eel fishing, and Gwin had deliberately included a "fish story" about seeing a huge eel in the river. Gwin decided the story had been too believable. Gwin, as a former reporter, knew the field—sooner or later, another reporter would hear the story. So he and Newton sat in the local cafe on St. Simons Island each morning, embellishing and fine-tuning the fish story into a narrative of folktale proportions. Should the opportunity arise, the next reporter would get a polished, detailed, and completely fabricated story.

That opportunity arrived in the guise of Jingle Davis, a reporter with *The Atlanta Constitution* who covered South Georgia and the coast. Davis, who had gotten her start locally at *The Darien News*, was a native of St. Simons Island and knew the river and marshes were home to a variety of rare animals. Whether she believed him or not, she reported the incident as a straight news story.

Gwin's tale began as he was eel fishing with Steven Wilson in late December on the banks of the river. From their small boat, they spied a snake-like creature 15-to-20 feet long, and as thick as a man's torso with two brownish humps about five feet across. It came disturbingly close to the small craft before submerging in a big swirl of water that rocked the boat. They watched as it swam away, leaving a swell like a boat wake.

It was supposed to be a joke, a giant eel yarn. But the joke was on Gwin—Davis had covered the waterways long enough to have run across other reports of odd encounters. None were newsworthy alone but added to Gwin's story, it became a series of sightings. Davis added a story from several years earlier upriver at Frank Culpepper's Two Way Fish Camp. Harvey Blackman had been at the camp, standing on a floating dock that was suddenly rocked by a large wave. When he looked for the source of the wave, something looked back that looked like a snake's head. Culpepper didn't see the beast but saw the wake bumping boats around and recalled one of the guests coming in for a rifle.

Davis's article appeared in *The Atlanta Constitution* on February 16, 1981, and the *Constitution*'s news service spread it across the state. *The Darien News* was caught unaware and chagrined that they had somehow missed a local sea serpent. Davis's former husband, Ervin, was a local graphic artist. When he dropped by the paper to deliver ad layouts, he was asked about the monster. Whether he was in on Gwin's gag or merely suspicious so the tale, his retort was "That's the Altamaha-Ha!" The newspaper thought it sounded Indian, so it became the name of the monster, although Ervin Davis likely meant it to emphasize the "ha-ha" syllables.

Several regional newspapers and media outlets mounted expeditions with no success. Additional reports of unidentified marine would come into the towns, which were then good-naturedly dismissed as being further sightings of the "Ha-Ha." The encounters then faded into local folklore until 1996 when Ann Richardson Davis self-published a children's book called *The Tale of the Altamaha Monster*. An innocuous tale of Chattan, a Scottish loch monster who follows his young friends when the family immigrates to the new settlement of Darien, Davis set up a website to promote the book and other projects. And one of the features on the site was a reader-contributed list of Altamaha-ha sightings.

The website's launch in 1997 was perfect timing. Internet search engines were coming into their own, and there had been a series of cable television episodes on the Loch Ness monster at the end of the year—any search for "sea serpent" was bound to encounter Davis's website. Even a cursory glance through the reports shows a proliferation of anecdotal stories, most of which are easily dismissed as native species misidentified through confirmation bias—species such as the shortnose sturgeon, the Atlantic sturgeon, West Indian manatees, and elephant seals. One species that is overlooked is the Alligator Gar. In his book on gars, Mark Spitzer feels this may not be an accidental oversight, since the gar would offer a probable identity over a cryptid, dispelling the cryptid mystery. This deliberate oversight applies to other cryptids, such as those of Lake Norman and Crooked Lake, discussed in the previous chapter.

In spite of Spitzer's caution, momentum was building for a cryptid origin. Karl Shuker briefly mentioned the monster and the website in his cryptozoology news column in the January 1998 issue of the *Fortean Times,* but it was a March 1998 column by syndicated folklore columnist E. Randall Floyd that reinvigorated interest in the Altamaha-ha. Relying on accounts from the Davis website, it raised the profile of the river beast again. *Fate* magazine picked up Floyd's column in their September issue, and suddenly these stories were considered a history, with unverified reports dating back to 1959. Davis also considered the schooner *Eagle's* 1830 encounter (Chapter 7) inside Saint Simons Bay to be an early Altamaha-ha sighting. Suddenly Larry Gwin's little fish tale was a sea monster with a 150-year record of sightings.

In 2009, the Darien-Mcintosh County Chamber and Visitor Center unveiled a 20-foot statue of the Altamaha-ha, now repositioned as a tourist-friendly cryptid named "Altie." Darien treats Altie as an unofficial mascot, with its image on murals and billboards. The result is both a whimsical presence in town and a crypto-tourism cottage industry. The monument in the Visitor Center in particular places Darien in a category with Lake Champlain, Loch Ness, and most recently, Cape Ann, elevating a sea monster into a regional symbol. Cryptozoologist Ulrich Magin suggests that such community expressions, as demonstrated by statues, museums, and souvenir tchotchkes, could be interpreted as a modern equivalent of an altar to a contemporary version of a *genius loci*, a guardian spirit of a place. Magin offers that this imbues a sense of identity and place that supersedes whether a cryptid is "real" or not, and is less important than the sense of belonging it offers. This means it does not matter if Larry Gwin meant his fish tale as a prank or if it was a tall-tale that exceeded expectations; it became a community-building emblem. But it should be noted that Gwin died in 2011, having lived long enough to see Altie evolve into a tourist icon. More telling is that even in his obituary, he referred to the sea monster simply as the "Ha-Ha."

### North Fork St. Lucie River

The 1975 spring fishing season had started off with an oddity—the rare appearance of white marlin in waters off Martin County, migrating to the Caribbean for the summer. Six had been caught at the end of April alone. Based on the number of other fish being caught in the local waters, the marlin may have been following the food source closer to shore.

The season took an odder turn a week later. The May 8 issue of the *Stuart News* reported that Joan Stoyanoff was claiming she had seen an unidentified 30-foot aquatic creature near her home on the Palm City side of the North Fork of the St. Lucie River. Mrs. Stoyanoff said she was fishing on her dock on April 9 with her neighbor JoAnne Heyer when all of a sudden, they heard "a blowing-gurgling sound."

"We turned around and saw between two pilings, which are about thirty feet apart, a brownish-grey object. At first I thought it was a manatee, but it was too long and big to be a manatee," Mrs. Stoyanoff reported. She added that the creature then sank back into the water and headed south toward Bessey Creek. Mrs. Heyer concurred that it was not a manatee.

There were other witnesses. Mrs. R. J. Haas, who lived along the same stretch of Murphy Road as the Heyers and Stoyanoffs, saw it from 40 feet away. She described it as looking "like a boat coming out of the water." Then, as it surfaced, she saw fish leaping into the air to escape being eaten, and water turbulence surrounding the creature.

Mrs. Addie Shaw, a housemaid for some of the Murphy Road houses, also saw the mystery beast in Mrs. Stoyanoff's yard. "It came out of the water and sank back down in the water...I've never seen anything like it around the waterway," she said.

The *Stuart News* contacted the University of Miami and spoke to Dr. Daniel Odell. He first suspected it was an odontocete whale—a whale with teeth for hunting and eating fish. This would suggest a sperm whale, a pilot whale, or a beaked whale. Odell noted a whale would have no competition from other whales in a river. Or it could be coming and going based on food sources. He also advised the readers that if it was a whale, it wasn't lost. With their sonar system, the whale was there because it wanted to be there.

Reporter Douglas Butler ended the article with the observation that there had been no further reports in the last week and a half. But "something is swimming the North Fork St. Lucie River that has the local residents gazing into the waters in search of something large and mysterious."

The mid-1970s were the height of interest in the Loch Ness monster, and the same researchers that had taken the famous Rines-Edgerton flippers photos three years earlier were back at the loch. If Butler had been trying to create a sea serpent sighting to siphon interest away from Scotland, it was a short-lived attempt.

Less than three weeks later, the *Stuart News* published a follow-up story. Joan Stoyanoff and her neighbors had seen the creature four more times and, having had time to read up on Dr. Odell's suggestion, now believed it was a 40-foot sperm whale. Stoyanoff was concerned that the whale was lost and wanted the poor creature towed back to sea. The Department of Natural Resources didn't believe her, let alone plan to assist the whale. Other residents had contacted marine biologist Bill Gehring at the Florida Institute of Technology in neighboring Jensen Beach. He would not commit to an identification without actually seeing the whale in question, but told the newspaper that towing whales back to open sea usually caused stress and injury to the whale, and was not typically necessary. Dr. Gehring also questioned if a sperm whale of that size would even be able to go into the shallow waters of the North Fork. It sounded more like a pilot whale to him, but he reminded *Stuart News* readers that if it was a sperm whale, it was an endangered species protected by federal and state law and should not be interfered with.

Had the whale waited a month or two before visiting, perhaps the newspaper could have piggybacked the misplaced whale on the rumors of new photographic evidence at Loch Ness. What is referred to as Nessie's "gargoyle headshot" wasn't made public until that winter, but rumors began circulating soon after the image was photographed in June. Alas, no one ever claimed whales had a flair for the dramatic.

# CHAPTER 11
## St. Johns Menagerie

Even describing Florida's St. Johns River is difficult. The river flows north, meaning downstate is upriver and downriver is to the north. The longest river in Florida, it winds 310 miles from its headwaters in the marshes of Indian River County, then turns to the east at Jacksonville before emptying into the Atlantic. Since it begins among the swamp serpents of the Treasure Coast and ends in the middle of an area of sea serpent sightings, it is only logical that the St. Johns also has a history of cryptids.

The river's elevation drops less than 30 feet over its length, giving it a slow-moving current. Jacksonville is close enough to the river's mouth that the incoming tide can force seawater up the river, briefly making the river reverse direction. The river is so slow-moving that it also has named lakes within the river, prompting the Seminóle name *Ylacca*, the "River of Lakes."

The result is 310 miles of very different environments—the upper St. Johns is marshy, the middle section is lakes and swamps, and the lower St. Johns is a brackish estuary. Add the variety of names assigned to sightings, and the river prompts cryptozoological confusion. Encounters are becoming increasingly homogeneous, in spite of taking place decades and miles apart and of different sizes, colors, and behaviors. If one is to read the current crop of folklore/cryptozoology books, The St. Johns Monster is 175 years old, big and gray, or serpentine and gray, when it isn't pink, and/or euryhalinic as necessary.

### William Bartram's Dragon

The newest addition to the cryptozoological lore of the St. Johns would also be the oldest, if any part of the story was remotely correct. Naturalist William Bartram made two trips up the St. Johns. His narrative is a combination of the two visits, which can be confusing since it has seasonal events happening at the wrong time of the year (for instance, a hurricane in May). Using his report to his patron, Dr. John Fothergill, instead of his actual 1791 book, it becomes easier to separate the two trips.

Bartram traveled the length of Lake George in May 1774. He had been unable to find a local guide, and instead, his small boat followed a larger trad-ing vessel. He soon realized why no Native would guide him when the ships were forced back to port by a gale. Finally reaching the far shore on the next attempt, he camped near modern-day Astor. The next day, he continued up-river (southward) to Idlewilde Point on the west bank of the river opposite the mouth of Lake Dexter. This is the mouth of Stagger Mud Lake, which Bartram referred to as "Battle Lagoon." Here he repeatedly battled aggressive alligators near 20 feet in length. In one widely reprinted account, these "river monsters" attacked the ship from all sides as Bartram fought them off with a club. Groups of them followed him to his camp where he finally drove them off by shooting at them with the gun he had neglected to put in the boat.

This adventure, while exciting, obviously does not involve a cryptid, other than Bartram, not without cause, referring to his reptilian adversaries as "river monsters." But in a 2012 recounting, Bartram had his Indian guide refusing to cross Lake George because of a "beast of the deep" and later seeing "fiery dragons" around Volusia Bar (where the St. Johns flows into Lake George). The source was a brief article on a paranormal blog site run by a Fortean re-searcher in Pennsylvania.

The article was by occasional contributor David Hoes, whose primary source was Count Albert Wass de Czege, the editor of the small newspaper in Astor, site of the 1953 river monster sightings. Wass also wrote an unpublished history of Astor, which included local monster lore. In addition to conflating Pinky with the Astor Monster, Wass completely misread Bartram's text. The passage in question refers to a hurricane Bartram was caught in, which he de-scribes, in his usual florid prose, how "the mighty cloud now expands its sable wings," and the wind drives it onward "spreading his livid wings" and spewing "fiery shafts of lightning." Compounding the error, the storm took place in Georgia, not on the St. Johns.

The article was noticed by Florida's self-proclaimed "Master of the Weird" Charlie Carlson, who was compiling notes for his next book. Carlson died in 2015, but he had shared the manuscript with Robert Robinson, who included the Bartram story in his 2016 book on paranormal tourism. It was only in 2017 that the story was widely noticed, when folklorist Mark Muncy, also drawing from Carlson's notes, included the new version of Bartram's journey in his popular book. Muncy's first book was successful enough that amateur paranormal websites began posting the story on the internet again. William Bartram may not have encountered a dragon, but he did assist in a cautionary tale about using unconfirmed sources online.

## Lake Monroe and other Confusions

In 1983, Loren Coleman released his seminal work, *Mysterious America*. An appendix in the back lists lake and river monsters of North America, a response to Roy Mackal's 1980 *Searching for Hidden Animals,* which includes an incomplete list of 30 cryptids in American waters, most of which were vague (Florida only shows the 1975 sighting on "Pinky" in Jacksonville).

In Coleman's book, Florida has five listings: Lake Clinch (Chapter 9), North Fork of the St. Lucie River (Chapter 10), Suwanee River (Chapter 13), The St. John's River, and Lake Monroe. This is where the confusion starts—Lake Monroe is one of the lakes in the St. Johns. The Florida list remained unchanged through the various editions of the book. By the time he and Huyghe released their *Field Guide to Lake Monsters,* the appendix included worldwide aquatic cryptids, but with the same five listings in Florida.

These five Florida listings, as well as the global list, have been extensively copied in books and on the internet, but no one, including Coleman, has ever elaborated on the monster supposedly in Lake Monroe. It remains a listing that no one has ever written about or found a source to cite.

The answer may be that Coleman, in the late 70s/early 80s, while compiling *Mysterious America* in the pre-computer days of book writing, mistakenly included Lake Monroe, Indiana, for the similarly named lake in Florida. The Indiana lake was the epicenter of a spate of regional bigfoot reports, both before and after the book's publication, so an errant note from the book's bigfoot section could have ended up in the lake monster section.

Another less likely alternative is that the Lake Monroe monster was a misnamed reference to the particularly obscure serpent report in 1905 that Coleman had learned of in his investigations. The *Tampa Tribune* had run an article about the Espiritu Santo Springs in Green Springs. The owners had just added bathing tanks and a pavilion to capitalize on the growing reputation of the mineral springs as a medicinal cure-all. The *Tribune's* local correspondent also filed a small article about a report of a 300-foot sea serpent spotted in Tampa Bay near the springs, joking that the spring waters must be more potent than expected. The *Pensacola Journal* picked up the article and spread the story even further. Another Green Springs resident took offense at what he considered fake news and threatened to punch the reporter in the face. Both papers published the hysterically angry letter. It generated a response from the newspaper, a slight increase in curiosity seekers, and then died off. In 1917, Green Springs officially incorporated as the City of Safety Harbor.

The only way this sighting works as the Lake Monroe monster is because there is a better-known mineral spring, also named Green Springs, which even has a history of medicinal water cures, in Deltona, on Lake Monroe. Now a state park built around the 76-foot-deep green sulfur spring, it would be easy for someone to confuse a reference to a Green Springs serpent now, a century after Green Springs became Safety Harbor, to Deltona's Green Springs.

It is not the only lake monster with a Florida residence that raises more questions about the sighting than the cryptid itself. The story begins in 2002 with an article by reporter Mike Archer about the Astor Monster in the *Orlando Sentinel*. In that piece, he mentions an 1896 account when a fishing boat and, a week later, a steamer, each had an encounter with a "beast of the deep" on Lake Dexter. At first glance, this is a reasonable claim, since Archer was interviewing Buck Dillard, who first saw the Astor Monster in Lake Dexter. The problem is that there is a significant difference between the draft of a steamship and that of a fishing boat, and Lake Dexter is a shallow lake. Per the NOAA charts, Lake Dexter is about four-to-six-feet deep, with submerged obstacles scattered across the bottom. Steamboats could not enter Lake Dexter. They passed the mouth but stayed in a river channel maintained for their safe passage.

Various cryptozoology books added conflicting details. Greg Jenkins suggests it was a relocated Clearwater Three-Toes (Chapter 13) and mistakenly cited Loren Coleman's website (lorencoleman.com) as his source. David Hoe and Mark Muncy, much as they did with the William Bartram dragon, relied on Charlie Carlson's unfinished manuscript, resulting in a full-blown but incorrect citation. The Carlson version tells that the *St. Francis Gazette* of August 24, 1896, reported that "something dangerously large" overturned a fishing boat in Lake Monroe. A week later, there was a report that the steamer *Osceola* was struck by something large near the same location.

There was no steamer named *Osceola* plying the waters of Florida in 1896. The Clyde St. Johns River Line ran between Jacksonville and Palatka, but the steamer *Osceola* didn't start service until 1914. Steamships with that name were sailing on the Mississippi River and in the Great Lakes, which makes sense when considering that *St. Francis Gazette* was an English language newspaper in Saint-François-du-Lac, Quebec. Located on the Saint-François River where it meets the St. Lawrence River, Saint-François-du-Lac is on the steam route between Quebec, Montréal, and Lake Ontario. So, whatever the steamer *Osceola* struck, it was not a Floridian cryptid.

## The Ocklawaha Serpent

Before a booming tourist trade began in the late 19th century, the route from Palatka, up the St. Johns to the Ocklawaha River and then to Silver Springs, was used for transportation of goods and brave passengers on a smaller scale. The problem was that standard-sized steamships were too big and had too deep a draft for travel on the fast-moving, twisting river that was notorious for shifting shallow spots and hairpin turns clogged with debris. From the end of the Civil War into the mid-1870s, three vessels traveled the Ocklawaha. These three vessels designed for the challenges of the river—squat with a blunt bow, limited tonnage, and a short smokestack—were not built for comfort nor speed and, fully laden, ran uncomfortably low in the water. They would become prototypes for a slightly larger and more maneuverable "Ocklawaha steamer," but the steamships were still a necessity, not a luxury, even as attempts continued to make the river more navigable.

In June 1871, the *Palatka Herald* ran an article about an encounter so spectacular that it was picked and run by newspaper internationally. A steamer was rounding Sackett's Point, where the Ocklawaha widens, when the little ship "suddenly encountered a sea of alligators floundering and splashing the waters in every direction." The ship's momentum carried it forward and the vessel suddenly found itself surrounded by alligators that attacked the vessel. The ship was low enough in the water that the passengers were forced to stop the reptiles from climbing aboard with firearms and hand-spikes, but the alligators attempting to board the small ship were the least of their problems. Other gators were attacking the hull, and planks were beginning to give way.

Several deckhands were dragged off the ship and not seen again. Just when it appeared inevitable that the ship would sink and leave the passengers at the mercy of the alligators, the crew received assistance from an unexpected source—a giant serpent that swam toward the fracas, then submerged with the alligators in pursuit. Soon, the point where the river widened was blocked with dead alligators, and the river ran red with blood. There was no further sign of the serpent, but atypically, the passengers hoped the snake had survived.

Aside from the size of a snake capable of taking on multiple alligators, the other item of note is that the huge serpent was nonchalantly identified as probably the same one spotted the previous autumn in the St. Johns at the Devil's Elbow near Palatka (now known as Horseshoe Point). The encounter would prove to be typical of St. Johns sightings—a cryptid with multiple reports, distinctly different from other groupings in the river, with a brief window of sightings in a limited geographic range.

**The Astor Monster**

Astor lies entirely within the boundaries of the Ocala National Forest, on the west side of the St. Johns between Lake George and Lake Dexter. The town's history dates back before recorded history; when Pedro Menéndez de Avilés, the founder of St. Augustine, reached the area, it was already home to a Timucua Village. Naturalist William Bartram and novelist Marjorie Kinnan Rawlings would also write about the area. In spite of this rich historical and literary legacy, it wasn't until 1953 that anyone mentioned a river monster.

The first people to report seeing a monster were Art and Gin Myers of St. Louis, Missouri. On Sunday, October 19, the couple had hired a fishing guide from the fishing camp where they were staying, Astor Cottages, owned and operated by Dan McNulty. The guide had taken them to a spot in Lake Dexter, a section of the St. Johns about four miles south of the Astor Bridge. They described the cryptid as having a head 30 inches wide with a 10-inch-long horn and a round nose "like a hippopotamus." The creature looked at the boat for a minute and then submerged and swam away.

The creature was next seen back near the Astor Bridge. A witness saw the cryptid swimming along the bank when the animal climbed up on the bank and disappeared into the forest. It had walked awkwardly, but it was walking on four legs. A third sighting took place near the bridge when a local woman was frightened by the creature sticking its head out of the water. Two additional sightings were reported in the *Astor News*. McNulty updated his notes to add that another fishing party had also seen the beast, bringing the count to seven separate reports since the first report, 24 hours earlier.

Within a week, the reports brought an influx of curiosity seekers and biologists, as well as the identity of the fishing guide who had brought the Myers to Lake Dexter. Buck Dillard had been the guide who brought the Myers to Lake Dexter, and he was incensed at armchair biologists who had dismissed the monster as a manatee. Dillard, having spent 35 years on the river, was adamant he knew a manatee when he saw one, and this was not a manatee. The denial that it was not a manatee would practically become Dillard's mantra—he was still repeating it in a 1982 interview, 30 years after the sighting and three months before his death.

One final sighting was reported at the end of the month, by a trio of bowhunters looking for deer, which provided the best description of the Astor Monster. Jack Miller of Jacksonville was entertaining guests, Arno Richardson and his daughter Mary Lou of Atlanta, Georgia. They had parked off Alco Road, where an unpaved National Forest Service access road forked off, and proceeded on foot east toward the St. Johns. They followed a deer trail for about a mile when they came to a swampy thicket. As Miller tried to push into the thicket, the brush began to sway and crack. Thinking they had found a deer, Miller pushed aside some bushes to get a clear shot. It was 20 feet away, but it was not a deer. He thought it was a sizeable rock at first. All he could tell is that it was big and it had a "bitter ugly smell."

When Mary Lou approached, Miller warned her to stop. At the sound of his voice, the beast began to run, making a terrible noise. Miss Richardson described it as "gray, clumsy, and huge," standing five-to-six feet at the shoulder and rapidly declining so that the tail was no more than a foot tall. It had a long thick neck and held its head much higher than its shoulders. It had a horn on its forehead that curved forward, and a cow-like snout and eyes.

Combining Mary Lou Richardson's description with Buck Dillard's insistence that it was not a manatee, there is a highly unlikely alternative: a southern elephant seal (*Mirounga leonina*). Although found primarily on the coast of Antarctica and the tip of South America, the male elephant seals are known to travel great distances to go ashore and molt all of their old fur and epidermal skin, a process that takes about a month, the same amount of time as the main sighting period of the Astor monster.

These massive animals are at home at sea and awkward on land, but they can move at speeds up to five miles per hour if threatened, by flexing their hind flippers and moving as if running on four legs. The largest southern elephant seal males weigh nearly 9 tons and are 20 feet long and have a large, inflatable nose that could explain the horn. But the real issue, as also discussed in the chapter on Mystery Seals (Chapter 12), is logistics. Wandering or vagrant males have been documented as far north as Ecuador on the west coast of South America, so there is no reason a vagrant couldn't make a similar 2,000 mile trip on the east side of South America. Such a position would put the vagrant in the North Atlantic Deep Water, the cold water current opposite of the Gulf Stream that could lead an errant pinniped to the Florida coast.

Countering this improbable journey, every marine biologist interviewed by the local newspapers was as adamant that it was a manatee as Buck Dillard was that it wasn't a manatee. Could they both be correct?

Ormund Powers was the only reporter covering Lake County for the *Orlando Sentinel,* and he had a lot of territory to cover with very few news items meriting coverage. So he relied on his own network of contacts for content, including the *Astor News*, a mimeographed weekly newspaper published by an exiled Hungarian nobleman who had a vested interest in keeping up the flow of hunters and curiosity seekers; weeks before the first sighting, Count Albert Wass de Czege had opened a restaurant in town. This is not to suggest Count Wass de Czege was fabricating reports, but as a local restauranteur, he would also be aware of how few of these new sightseers had ever seen a manatee, and how the Astor Monster sightings overlapped with the annual manatee migration. From November to March, hundreds of manatees seek shelter from the cold by proceeding up the St. Johns to the 74° waters of Blue Spring, the largest spring on the river and now a state park and designated Manatee Refuge. It is also 20 miles upriver from Aston, giving novice monster-hunters ample aquatic creatures to spy upon as they passed through town. On November 21, the *Orlando Sentinel,* quoting *Astor News,* notes that although the monster had not been seen in three weeks, footprints had been found in the mud on a small island near Lake Dexter. Manatees are known to "pec walk," a rarely observed behavior where the manatee partially leaves the water to graze on the plants, leaving behind tracks among the crushed brush and chewed upon weeds.

Paranormal researcher Charlie Carlson paints the St. Johns River in such broad strokes that he covers the history of cryptid sightings in one page, blurring distinctly separate encounters together and glossing over others entirely. Although his Astor Monster reference misleads by making the encounters range between Astor and Blue Springs, he does include illustrations from witnesses, including one of a "brontosaurus, but not nearly as large" creature that is unquestionably a pec-walking manatee grazing on the riverbank.

Regardless of the identity of the Astor Monster, it was far from the oddest cryptid reported in the river and certainly not the most colorful.

**Pinky and Company**

If ever there was a cryptid adept at muddying the waters both figuratively and literally, it was the cryptid spotted on May 10, 1975, during a fishing trip about five miles south of Jacksonville. Charles Abram, president of Harrell Glass and Supply Co, and his wife, Dorothy, had taken friend Brenda Langley, and employees Wallace McLean and Edward St. John out for an early morning fishing trip. Launching his outboard boat from the docks at Clapboard Creek Fish Camp, they traveled about five miles to Abram's preferred fishing spot, along Quarantine Island, approximately where I-295 now across the river. By 10 a.m., it was apparent there was a storm rolling in, so Abram decided to head back. Suddenly Brenda Langley saw something raise its head out of the water 50 feet from the boat. To her growing concern, it came closer; she thought that it was unafraid, perhaps even curious about the boat. She yelled for the others to look, but it had submerged by the time they turned around, distracted by the storm clouds. But it came back up 20 feet away for eight seconds, which was just enough time for all five to view it as the boat turned toward home. Dorothy Abram described it as like a "dinosaur with its skin pulled back so all its bones were showing."

Brenda Langley elaborated on the description, noting the "head was as big or bigger than the head of a person, it had two horns on its head like snail horns, little horns or fins or ridges down the back of its long neck, a mouth that turned down at the edge and either gills or flaps on each side of the mouth." She also added that it had "big, dark, slanted eyes" and was so ugly looking that she thought it looked like "pictures you see of dragons." Both women agreed on one detail: the creature was pink, "the color of boiled shrimp."

The boat made a hasty retreat, and not just because of the oncoming storm. All five passengers saw the creature but never saw a body, only the neck and head. After the shock wore off, Dorothy Abram confided the sighting to a friend, who immediately called the Jacksonville paper. *The Florida Times-Union* went looking for the Abrams, who admitted they didn't report it because they knew no one would believe them.

The *Times-Union* ran the article on May 16, which was then picked up by UPI for the wire service. By May 17, the story of the "scrawny-looking, pink river monster" had gone national. Locally, a follow-up article noted that when Mrs. Abram was asked to compare her recollection to pictures of types of sturgeons, having been told the river monsters were usually sturgeon, said she was not convinced. In addition to sturgeon not having necks to raise their heads out of the water, they were most assuredly not pink.

On June 8, Alton Slagle of the *New York Daily News* devoted his syndicated column to the pink monster, light-heartedly noting the appearance coincided with the start of the annual monster sighting season in Florida. The legend of Pinky was born.

Even though Slagle's tongue was firmly planted in his cheek, his journalistic gravitas as a hard news and science reporter gave national legitimacy to the name Pinky and gave Pinky a profile so high that when additional sightings began to appear at the end of 1975, they were automatically grouped with Pinky, in spite of the fact the descriptions were unquestionably not the same creature, let alone the same color.

A December 14 sighting by John Bomgardner described an entirely different creature. Bomgardner was on lunch break near the seawall when he heard something go "pssssh" and spray came up. Then, 100 feet from the shore, he spotted a "shiny black head ('the size of a good-sized watermelon') on a serpentine neck rose out of the water." It took a breath of air, and then a hump came up, like a snake. A coworker saw a split tail break the water after the foot-wide body submerged. Bomgardner thought it was about eight-feet long, but a coworker who also saw the creature thought it was closer to 20-feet long.

Bomgardner's description of the sound of venting spray is indicative of a marine mammal surfacing to breathe, most likely a pinniped. Zoologists Woodley, Naish, and Shanahan published an overview of the potential of undiscovered pinnipeds still remaining. In discussing Heuvelman's long-necked seal theory, the zoologists suspect most reports are actually of a "strange, unfamiliar and long-necked animal" (at least to people only familiar with seals): the California Sea Lion (*Zalophus Californianus*). Sea lions continue to be sighted along the Florida coast. F. G. Wood of St. Augustine giant octopus fame (Chapter 6) and a team from Marineland captured a juvenile male California sea lion on Anastasia Island at Crescent Beach in the late 1950s, and by the late 1960s, marine biologist Gordon Gunter found so many reports that he believed California sea lions could survive for months in the warm waters of the Atlantic coast and that they could establish self-sustaining populations.

At approximately the same time that Bomgardner was seeing the creature in the river from the sea wall, Dave Green was crossing the nearby Fuller Warren Bridge. He noticed something in the water that he described it as quill-feathered and fan-tailed, like an eel with a ridged hump down the middle. It should be noted that Green's observation was made in a moving vehicle on a bridge 75 feet above the river.

In January 1976, *Tampa Tribune* reporter Jaclyn Dalrymple interviewed Wallace McLean and Ed St. John, the two remaining Pinky witnesses who had never talked to the press, adding their details. St. John thought the head was smaller than Langley had described it, "slightly larger than a football, squared off at the edges." He also thought the horns might have been ears. McLean sketched the creature, making the ridges that Langley described looking more like the bony plates on a stegosaurus. McLean described two gills on the lower part of the face, and the creature was "all wrinkled, like tired, old skin." Already, seven months after the sighting, there are notable differences in the description of the head and the neck, a phenomenon that Dr. Charles G. M. Paxton of the University of St Andrews refers to as the drift in testimony. Paxton, who uses statistical analysis in cryptozoology and marine sciences, finds that such a tendency for the story to begin changing is not unusual. Dalrymple then spoke to a skeptical Marine Patrol before talking to John Bomgardner about his December 14 sighting. Bomgardner reiterated his sighting, which just reinforced how distinctly different the creature he saw was from Pinky. Dave Green had changed his story, deciding it was a manatee, or whatever the Marine Patrol said it was. He was tired of the ridicule and impact on his HVAC company's business.

The article suggests a correlation was coalescing, grouping cryptids occurring six months apart in the same stretch of the St. Johns as one creature. Over the ensuing decades, subsequent sightings of unknown creatures were tacked on to suggest one cryptid collectively known as the St. Johns Monster or Pinky. When an April 1978 report came in of a 20-foot sea serpent north of the Crescent Beach Bridge in the Intracoastal Waterway, it was still the St. Johns monster regardless of the fact it was nine miles south of Saint Augustine, not in the St. Johns River. Kelly Parrish, the fisherman who saw the creature in the Intracoastal, described it as a huge snake that surfaced 75 feet from his boat. Like Bomgardner, Parrish's attention was caught by the beast expelling air. The creature undulated like a snake and was feeding, lowering its head into the water and then raising it with a mouthful of grass. Although Parrish insisted it was a serpent, the description of air-expulsion and grass-eating sounds similar to several manatees feeding in close formation.

The sightings around Jacksonville continued. In July of 1978, 20-30 people watched a creature unhurriedly feeding in the river 50 feet off Stockton Park, south of downtown Jacksonville in the Ortega neighborhood. It was described as a giant black snake with a head the size of a basketball, strikingly similar to Bomgardner's description of his sighting in 1975.

In 1989, reporter William Marden looked back at the sea serpent sightings in Jacksonville. (Loch Ness had been generating interest after Lowrance Electronics commercially released a videocassette of "Operation Deepscan," their large-scale sonar hunt for Loch Ness.) Marden interviewed John Bomgardner and Kelly Parrish about their sightings. Over a decade later, both still believe they had seen a sea serpent, and both frequently returned to the spot they saw the creature, camera at the ready. Both retellings showed evidence of Paxton's expected "drift in testimony." Bomgardner now stated he had seen the breath expelled from the creature's nose, and Parrish's serpent had gained an additional ten feet in length.

In 1993, Pamela Hicks, as was her habit, was looking out of her high-rise apartment in Jacksonville's Arlington neighborhood, across the St. Johns from downtown. As she watched the calm river, a 15-foot serpent surfaced. Hicks described it as dark gray with a long, skinny neck. The first thing it reminded her of was the "surgeon's photo" of the Loch Ness Monster. This sighting, as well as the one from 1978, was more reminiscent of encounters at the mouth of the river, such as the one reported by the steam packet *William Gaston* in 1853 (Chapter 7).

In 1997, looking back at the start of pollution reduction in the St. Johns, columnist Bill Foley, who, regardless of accuracy, seemed to enjoy pointing out the color of the local river monster, noted that pollution control had only been partially successful, but it did bring the return of the "regulation hump-backed, flat-snouted, big old sea monster with the defining characteristic of being the color of bubble gum." Because he felt "St. Johns Monster" was too serious a name for a pink monster, he was now referring to the creature as "St. Johns Johnny." It did not catch on.

In 2002, George Eberhart released his popular *Mysterious Creatures: A Guide to Cryptozoology.* The entry for Pinky in that book includes a bibliographic reference to an Ivan T. Sanderson article on "The Five Weirdest Wonders in the World" in a 1968 issue of *Argosy*, which discussed the Astor monster sightings in 1953 and his correspondence with Mary Lou Richardson. Sanderson refers to the Astor cryptid as the "St. Johns Monster." As a result of the Eberhart inclusion, the Sanderson article connects Pinky, already laden with unrelated sightings, with the 1953 sightings, in spite of the article being written seven years before Pinky was sighted.

In 2002, Mark Hall, a cryptozoologist inclined to assume any non-hominid cryptid was a surviving dinosaur, devoted an issue of his research journal *Wonders* to the colorful cryptid. He felt that Pinky was a dinosaur, specifically *Thescelosaurus neglectus*, a small herbivorous ornithopod that would have been alive at the time of the end of the Cretaceous. The locations where the fossil remains are found suggest that it may have preferred to live near water.

Hall's identification was flawed from the start. He adds horns to his version to more closely match the description. He admitted that horns are not a feature of the species, but suggested it is an adaptation because the thescelosaurus would need to become primarily aquatic for his theory to work, and therefore, the horns were actually erectile breathing tubes. Although he bases much of his identification on the description of Pinky by Dorothy Abram, its size comes from Lake Asor, via Ivan Sanderson's 1968 *Argosy* article.

The confusion became so firmly solidified that even Loren Coleman, one of the world's leading cryptozoologists, was misled. In 2008, he flew to Florida on a research expedition to look for material on Pinky, which he believed was the pink ornithopod dinosaur as suggested by Mark Hall in 1992. The problem was he was looking for "the unique river monsters seen up and down the St. Johns River in Florida, now known by the collective name most popularized in the area since the 1970s as 'Pinky.'" Because he had been led to believe that all St. John sightings were of Pinky, he started at Lake Monroe, 150 miles upriver from where Pinky had been spotted.

By the third day of his expedition, he realized the draft of the river made any sea serpent or dinosaur theory, such as Mark Hall's *Thescelosaurus*, untenable, but he remained focused on Lake Monroe. On the fifth day of the trip, he visited the Historical Society of Sanford. The director was unfamiliar with Pinky or any bipedal monsters in the area. So they introduced Coleman to Charlie Carlson, whose 1997 book had begun the process of evolving the Astor monster sightings to a more all-encompassing "St. Johns Monster." Carlson explained to Coleman that "Pinky" was used by cryptozoologists, not locals. Carlson was technically correct in that Pinky was not in local use. What he neglected to mention was the reason: Pinky wasn't a local cryptid. Carlson was also incorrect in claiming that cryptozoologists had coined the Pinky monicker. That dubious honor goes to Alton Slagle and the national distribution of his syndicated column.

Carlson's next book, in 2005, would differentiate between the 1975 sighting and the sightings along the river that preceded it, but it also emphasizes the "dragonlike" description and doesn't mention the pink color of the creature. Carlson then equates this generically colored Pinky with the probably fictitious Pablo Beach sea monster (Chapter 15).

Zoologist Karl Shuker has no use for Hall's pink dinosaur. He points out that Hall himself, in his 1991 book on monster lizards, researched pink lizards with horns like a cow or a moose that were supposedly three-to-eight-feet long, reported in "Catlick Creek Valley," which Hall has identified as Scippo Creek in Pickaway County, Ohio. Sometime before 1820, there was a drought that dried out their aquatic habitat, and a forest fire finished off any survivors. Shuker suggests that if the reported horns were equated with branching external gills, such as those found on the much smaller axolotl (*Ambystoma mexicanum*) of Mexico, Hall's theory that the pink lizards could offer a larval form of a salamander, significantly bigger than any known North American species.

Unfortunately, Shuker's focus on the horns/gills issue overlooks the fact Hall never tracked down his source for the story. The source material is actually a folktale about giant pink, fire-breathing lizards, which are only seen by the men who dared to go out at night in Salt Creek Valley, which storyteller

Erasmus Foster Darby (folklorist David K. Webb) states is the location, not Scippo Creek as Hall believed. As an example of how serious the actual tale was intended to be taken, the fire that burned the valley to the ground also took out the church and all 36 distilleries, after which the giant pink lizards were never seen again. In other words, it was a tall tale, not an early cryptid account.

Shuker is on firmer ground source-wise with a pink cryptid report from 1928 South Carolina. Herbert Ravenal Sass was a newspaper reporter turned nature essayist, writing about his beloved low county around Charleston, South Carolina. In 1928, Sass and his wife were in a flat-bottomed boat exploring the floodwaters from Goose Creek Lagoon north of Charleston, South Carolina. They were in an area near the ruins of the colonial Izard plantation house known as "The Elms." The estate covered the area from what is now the home of Charleston Southern University down to Goose Lake Lagoon.

Sass and his wife spotted something "moving through the water growths below the surface an indeterminate shape." Counterintuitively for a naturalist in shallow waters shared by alligators, Sass slid a paddle beneath the creature and lifted part of it above the water. "It was very heavy, about the thickness of a man's lower thigh, of a bright salmon pink and orange color. How long it was I don't know, because both ends remained underwater. This section was all we saw, and we only saw it for an instant before it slipped off the oar and vanished." Both Sass and his wife saw a pair of legs, "like an alligator's or salamander's." Sass's impression was that it was like a hellbender (*Cryptobranchus alleganiensis*) only far larger than the hellbender's average size (up to 29 inches).

Whatever it was, it was "so fantastically impossible" that Sass chose never to write about it. But he was insistent it was not an alligator, an animal with which he was very familiar. He closed the article by noting that it "may be even more fantastic to suggest that there exists in these Carolinian swamps and lagoons a species of giant amphibian of which no specimen has yet been taken."

Shuker notes Sass's impression of an oversized hellbender, which would need to be minimally five-to-six-feet long to match his description, sounds similar to Pinky. The description of Pinky and that of a hellbender have similarities: a large, flattened head, a mouth that is turned down at the edges, loose folds of skin on its neck that could be mistaken for gills, and wrinkles along the body that could conceivably be mistaken for protruding bones, although Wallace McLean's description of Pinkly being "all wrinkled, like tired, old skin" is even a closer match.

The one issue with a giant hellbender explanation, aside from the color, is Pinky's "horns." These horns could be explained as an external gill structure, the hellbender has no such structures—it loses its external gills when it reaches adulthood. Shuker speculates that these animals may be identified as giant hellbenders in a neotenic state. Neotenic amphibians never complete metamorphosis, keeping juvenile features in the adult animal, such as the axolotl with its external gills. Noting that if neotenic salamanders are lifted out of the water, their gills flatten against their neck and would not appear to be horns, Shuker borrows a page from Hall and suggests Pinky's horns are breathing tubes. Shuker remains highly skeptical of a living dinosaur but believes a giant salamander with breathing tubes and Pinky's prominent eyes, would be a new, unknown species and is far more likely than a dinosaur.

All of this for a pink cryptid sighted by five people once for less than 20 seconds.

# CHAPTER 12
## Mystery Seals

When asked about the Mansi photo of the Lake Champlain Monster at a cryptozoology conference in 1982, Forrest G. Wood of St. Augustine Giant Octopus fame famously remarked that just because it looks like a plesiosaur doesn't mean it is a plesiosaur. The long-necked seal is a hypothetical megapinniped, an oversized seal that accounts for, depending on which cryptozoologist you ask, a few, some, or most plesiosaur-like sea serpent sightings.

A.C. Oudemans, in *The Great Sea-Serpent*, suggested that a giant seal was the source of all sea serpent sightings, classifying it with the taxonomic name of *Megophias megophias*. Oudemans proposed this long-necked, long-tailed seal had diverged from all other pinnipeds early in pinniped evolution. Other characteristics included its massive size, up to 200 feet long, and a noticeable mane on the male.

Naturally, Bernard Heuvelmans disagreed, suggesting *Megophias* was a composite of several different sea serpents. Heuvelmans' book was an attempt to retrofit historical sightings from across the globe into nine distinctive categories. Several contemporary zoologists, most notably Ulrich Magin in 1996 and Darren Naish in 2001, point out that even Heuvelmans couldn't keep his classifications straight.

Instead of Oudemans' megapinniped, Heuvelmans suggested a megapinniped, but a different one. His creature was also a giant, long-neck seal, but without the long tail. He gave it an alternative classification, *Megalotaria longicollis*. According to Heuvelmans, *Megalotaria* evolved to fill the ecological niche left vacant by the extinction of plesiosaurs in the Late Cretaceous. This convergent evolution is why the mammal was often mistaken for a plesiosaur. *Megalotaria* would have had limited mobility on dry land and spent most of its time in the water. Sightings were scarce because this proposed animal had other specialized adaptations such as its nostrils become snorkel, allowing it to breathe without fully surfacing.

Heuvelmans further suggested the creature was also euryhaline—able to adapt to a wide range of water salinity, which conveniently allowed the same identification to be applied to lake monsters of this type. Irish historian Peter Costello, in particular, ran with this concept. His *In Search of Lake Monsters* uses the long-necked seal as the identity for every lake cryptid from Ogopogo to Loch Ness. In *The Monsters of Loch Ness,* biologist Roy Mackal doesn't dismiss the euryhalinic theory, noting that there are plesiosaur fossils that were found in contexts of freshwater environments.

Heuvelmans also had a separate category for the "merhorse" of classic mythology. He considered it another pinniped with a long neck, but with more of a pronounced "horse head" and a mane as a male gender trait. Subsequent attempts at a marine cryptid classification system, such as Coleman and Huyghe's *Field Guide to Lake Monsters, Sea Serpents, and Other Mystery Denizens of the Deep*, merged merhorses and long-necks back into a single sea serpent type, which they dubbed the waterhorse.

Roy Mackal, trying to identify the Loch Ness monster, considered a marine mammal first, before deciding on a large, previously unknown amphibian. The marine mammal he considered most suitable was a sirenian, not a pinniped—the long-extinct Steller's sea cow (*Rhytina stelleri*). Mackal acknowledged that sirenians (manatees and dugongs) all have very short necks, but since the cervical vertebrae in these animals are not fused, it would be possible that a long-necked, small-headed form of sirenian could have evolved.

Christian de Muizon of the Muséum National d'Histoire Naturelle in Paris did a study of fossils found along the coast of Peru, including the most complete remains of *Acrophoca longirostris*, the swan-necked seal. An offshoot that went extinct in the late Miocene (11 to 5 million years ago), it is most closely related to the highly adapted lobodontine seals. It is interesting to note that the fossil record indicates a highly flexible spine that, in addition to a wide range of motion in the neck, suggests *Acrophoca* swam with a horizontal motion like a serpent. The other notable modification is that the skull indicates a lengthening of the snout to a degree not found in any other current or fossil seal.

*Acrophoca*—with its long, flexible neck, serpentine swimming style, and elongated (horse-like?) snout—does initially appear a promising candidate. But the swan-neck seal's neck is not nearly as long as needed for the proposed long-neck megapinniped, and there is no fossil record to suggest it continued to evolve a longer one. De Muizon suggests the fossil physiology indicates it was probably not as good a swimmer and diver as today's seals, swimming slower and staying underwater for less time. The associated fauna in the fossil record suggests it stayed in the shallow waters along the shore and was, therefore, a more coastal seal than the modern lobodontines. Later swan-necked seal fossils uncovered in Chile by Walsh and Naish suggest that *Acrophoca*, or at least the Chilean version, may have actually been better suited for swimming in open waters than de Muizon suggests.

Pelagic or littoral, size discrepancies eliminate the swan-necked seal as Oudemans's one universal sea serpent. Assuming a long-necked pinniped existed into historical times, humanity may have inadvertently killed off the species. A. M. Springer, looking at the precipitous decline of seals, sea lions, and sea otters in the North Pacific, theorizes the collapse was caused by increased predation by killer whales. Killer whales, foremost natural predators of great whales, were driven to feed more aggressively on smaller marine mammals because the great whales were overhunted. The long-necked seal may also have been a victim of the killer whales as they worked their way into progressively smaller food species.

### The White River Monster

In 1937, Arkansas's Jackson County was farming country, centered around the county seat of Newport. And the lifeblood of this agricultural economy was the White River, providing access to the Mississippi River and the major southern ports. Commercial fishing, trapping, and lumbering along the banks provided income supplements to the farmers. Newport was also a national center for button making, which meant it was possible to make a seasonal living at shelling, harvesting river mussels for their nacre, the iridescent inner shells, which were made into mother-of-pearl buttons.

Everything changed in June of that year. About six miles south of Newport, on the west bank of White River in the township of Bateman, Silvia Wyatt, a black sharecropper, happened to glance across the yard at the White River. There was something in the river. Her husband Dee was summoned and also witnessed a large and unfamiliar creature in the water. Heading up the road to the home of Bramlett Bateman, their landlord, they told him they had just seen a monster. Before Bateman could ask the obvious question, they insisted neither of them had been drinking.

The dubious Bateman followed them back to their farm and immediately became a believer. Bateman would later describe it as "as big as a boxcar." Bateman went to town and informed the Chamber of Commerce about the monster. Twelve "reputable citizens" then came and witnessed the beast. Bateman fenced in the viewing area, started charging 25¢ admission, and hired local teens to run a concession stand. Signs were placed on the major roads that pointed the direction to the "White River Monster." The location was a mile-long stretch of deep (75 feet), quiet water that did not have an official name, in spite of Bramlett Bateman's best efforts to refer to the area at "Bateman's Eddy."

As word spread, so did media coverage, courtesy of a slow news period. The Associated Press newswire covered new sightings and reports of the occasional burst of bubbles from the bottom. By July, those theories included a giant sturgeon, a manatee, an alligator gar, a hidden moonshine tank, and a sunken scow that occasionally floated to the surface, buoyed by gases from decaying plant matter beneath it.

On July 13, W. E. Penix, a local toll-bridge collector, announced a plan to identify the creature. He canvassed the area for spare rope, enough to make a fishing net 40 feet long, 15 feet wide, and with six-to-inch-inch meshes. Penix didn't expect to capture the monster, merely entangle it long enough to identify the species. The attempt ran out of rope and money before it came to fruition.

By then, the Chamber of Commerce had seen the publicity potential and joined the search. They announced they had hired a diver to descend to the bottom of the river and search for the monster. Former Navy diver Charles Brown would address reports that the Chamber had invented the creature as a publicity stunt. The plan was for Brown to dive for four days. He expected an oversized catfish but did carry an eight-foot harpoon, just in case.

The event opened with great fanfare. The chamber took over admissions, and Bateman sold barbecued goat sandwiches and cold drinks to hundreds of out-of-towners. A loudspeaker on a barge in the middle of the river broadcast the diver's observations. Unfortunately, his only observation was that the river was muddier than usual because of heavy rains and his visibility was under three inches from his helmet. A second dive in the afternoon revealed no secrets because Brown had to be pulled out of the mud on the bottom. However, the spectators danced well into the night on a temporary dance floor to music broadcasting from the loudspeaker on the barge.

The next day was even worse. A valve jammed in Brown's helmet, and the diver became a balloon bobbing on the surface. By the time the suit was repaired, the crowds had begun to drift away. Brown didn't bother diving again. Even an article in *Time* magazine couldn't revive interest. Locals continued to report sightings, but the moment had passed. Bateman would later recollect he saw "Whitey" well into 1938 before it vanished.

In 1940, the monster made a brief reappearance. A nine-foot, 230-pound gar was caught in the Old River, a tributary of the White River, and proclaimed to be the monster. The victory was short-lived. D. N. Graves, the secretary of the game and fish commission, countered the claim by noting a story told to him by a friend of a fisherman who claimed he created the monster. This friend of a friend says he came across a bed of yellow sand-shell mussels (*Lampsilis anodontoides*), the most valuable of all commercial shells. Too valuable for button making, the shells were exported for use in knife handles, earrings, and novelties, anything requiring a long, straight piece of mother-of-pearl. Realizing his bounty, the fisherman rigged an old scow with wires, flipped over, and submerged it. He then started the monster rumors, and when someone came upriver to investigate, he'd raise the boat from the depths, frightening off any potential competition for his lucrative mussels. The story was obviously invented and didn't match the events of 1937, but the tale subsequently sank any renewed interest as quickly as the monster.

Modern sightings of the White River Monster started with a front-page article in the June 18, 1971, issue of the *Newport Daily Independent*. The timing of the river monster's reappearance was noted by the media—who were collectively more skeptical than their 1937 peers—less than a month after the Associated Press propelled the Fouke Monster, a hominid cryptid on the other side of the state, into national awareness. An anonymous call to the newspaper triggered the White River Monster article. Mike Masterson was the editor of the *Independent*. Masterson was not only new in town, but it was also his first newspaper job after graduating from the University of Central Arkansas earlier that year. Whether he received the call as a prank and saw the 1937 articles in the newspaper's morgue as confirmation, or if he fabricated the call to create a story, as some residents later suggested, the story was timed perfectly. In addition to the Fouke Monster, the creatures in both Loch Ness and Lake Champlain had been increasingly in the news. And then the photograph appeared.

On June 28, Cloyce Warren and two friends were fishing in a boat near the Newport Bridge. The men watched something bubble to the surface and swim away from the boat. Warren had brought a Polaroid camera to shoot pictures of their catches. He was able to take one shot before the creature submerged again. He recalled it to be gray, 30-to-40 feet long, and having a spiny ridge along the spine. The picture, in the proud tradition of cryptids, is blurry and inconclusive but seems to show something in the river—a mass that could be organic, or it could be rocks breaking the surface. There is a line trailing away from the mass that could be a tail or a mostly submerged spiny back, but skeptics such as Joe Nickell note the line is lit differently, and he suspects the line is merely a series of dashes added with a pen after the photo was developed and before publication.

The White River of 1971 was not the same river as the White River of 1937. With the river more accessible and now heavily used for recreation, claims of sightings and encounters were not limited to the stretch of water near the Bateman Farm. On July 5, 1971, the Jackson County Sheriff was called to investigate strange footprints found on Towhead Island, just north of the Bateman farm. The tracks were 14-inches long and eight-inches wide, with three long toes or claws and possibly a spur projecting from the heel. Bent saplings and crushed underbrush indicated that something large creature had been on the island.

The final sighting of the year was made by 63-year-old farmer Ollie Ritcheson and his grandson Joey Dupree. Ritcheson took a small boat down-river, planning to look at the tracks on Towhead Island (Ritcheson initially claimed they were going fishing for catfish). As they passed the 1937 sighting area near Bateman's farm, the boat suddenly struck something in the middle of the river—elevating the craft and terrifying the men. They were convinced they had struck Whitey, which was attempting to overturn the vessel. More likely, the two had hit a newly created natural levee, formed by silt deposited by the river as the flow slowed down over the deep-water channel that had been the primary Bateman sighting location.

CBS and Nippon TV both sent film crews to Newport. Bramlett Bateman had reached his fill and refused to cooperate with either film crew. The rest of the town, however, was more than happy to take up the slack. By now, the monster was a marketing commodity, and curiosity seekers would not be limited to watching a diver and eating barbecue goat sandwiches. Images of a generic sea monster, regardless of similarities to actual reports, appeared on souvenirs, in advertising, even on the cover of the local phonebook.

This time, the White River Monster would not fade back into obscurity, thanks to the TV coverage and growing interest in cryptozoology and a re-surgence in folklore. In 1972, the Loch Ness Investigation Bureau captured underwater images that appeared to be the monster's flipper, and Arkansas state Senator Robert Harvey responded by sponsoring legislation creating the "White River Monster Sanctuary and Retreat," creating a state-protected game preserve for the creature. Under state law, it was now illegal to harm the monster if it is found in the river between Old Grand Glaise and Rosie, creating a 20 mile stretch of river, roughly ten miles upriver and downstream of New-port (including Towhead Island) prohibiting anyone from "molesting, killing, trampling, or harming" the creature.

About that time, William Harris, a history teacher in nearby Greer's Ferry, began actively researching the legends of eastern Arkansas, which he considered neglected in folklore research circles, particularly the body of lore among the minority sharecroppers and transient farmworkers.

A native of Newport, Harris grew up with stories of the White River Monster. As part of the project, he decided to approach the locals and see how the 1937 and 1971 of sightings compared. To his delight, he found several residents still in town who recalled the 1937 events, including Bramlett Bateman himself. The general consensus was that those who remembered the 1937 events believed there had been a creature in the river, but most residents considered the 1971 return to have been started and maintained by the *Newport Daily Independent* as a marketing ploy. This is not to say the subsequent reports were fabrications, but the media coverage did seem to give witnesses the nerve to report their encounters. One person Harris talked to saw more than two creatures the size of cows at a nearby sandbar in 1974. He saw no head or tail, just backs roughly 5-to-6 feet long. He too had seen tracks "as big as your hand" along the riverbank where the weeds had been chewed up.

Harris further observed that the White River Monster fell within the general tradition of sea monster folklore Heuvelmans had noted in an early chapter of *In the Wake of the Sea-Serpents*. Heuvelmans believed sea serpent reports usually appeared during the summer months as a replacement for the lack of other news, adroitly remarking that in the summer months, "the serpent is more often found in the popular press than in the ocean."

Harris concurred. He noted that, in spite of newspaper claims of earlier sightings, an extensive review of local media, town histories, and records found no reference in print before 1937 that could be construed as a reference to the monster. Whatever Silvia Wyatt first saw and Bramlett Bateman then ran with, 1937 was the first encounter with Whitey. Harris also observed that the volume of articles in regional newspapers peaked in the summers of 1937 and 1971, then trailed off by September. He concluded that Whitey continued to appear each summer in both oral and popular contexts, but media appearances were less frequent each successive year. This may have been true in Harris's final report in 1977, but the White River Monster was already a regular fixture in cryptozoology circles.

Ivan T. Sanderson's Society for the Investigation of the Unexplained (SITU) was already aware of the reports as early as June 1971, but what caught the Society's attention were the reports of three-toed footprints. To them, this was a potential connection to the Florida Three-Toes sightings and Sanderson's giant penguin theory (Chapter 13). Their report in the October 1971 issue of *Pursuit* found too many variations in the stories to construct a concise description. Even eliminating the reports of the creature having a spiny-back as alligators (*Alligator mississippiensis*) do, the reports varied so much in size and anatomy as to be inconclusive. No definitive identification was offered, but giant penguins remained a default option.

Roy Mackal was more interested in how Whitey's appearance on the surface was preceded by surface disruptions and the release of bubbles that spread out in a ring. He also noted the discrepancies; sizes which ranged from 10-to-12 feet long to 60 feet long and skin that ranged from smooth to peeling to "crusted." Interestingly, Mackal also focused on one report of the head that claimed a horn was present. On that basis, he was definite in his identification: Whitey was not a cryptid, merely a misplaced but known animal. Specifically, Mackal identified the White River Monster as a large male elephant seal, but he couldn't specify whether it was a southern elephant seal (*Mirounga leonina*) or a northern elephant seal (*Mirounga angustirostris*), as either would fit since the two species are the largest extant non-cetacean marine mammals.

But as Joe Nickell notes in his *Skeptical Inquirer* article, the real issue with Mackal's identification is logistics. For Mackal's theory to be correct, two separate elephant seals would need to make the voyage, one in 1937 and another in 1971. Further making an elephant seal identification unlikely, only northern elephant seals, which inhabit the eastern Pacific Ocean, molt in the northern hemisphere's summer. Southern elephant seals, found primarily along the coast of Antarctica and the tip of South America, molt in the southern hemisphere's summer. So, not only would these two errant elephants have to swim over 425 miles up the Mississippi River from the Gulf of Mexico, then both turn at the White River and continue another 150 miles up to Newport, these massive pinnipeds would have to somehow crossed from the Pacific to the Atlantic and then across the Gulf of Mexico. Once would be highly unlikely, twice was statistically impossible.

Nickell suggests a more likely candidate—a Florida manatee (*Trichechus manatus latirostris*). Manatees can reach lengths of more than 14 feet and can weigh more than 3,000 pounds. Manatees are typically greyish brown in color but can also be covered in algae or a type of barnacles, explaining the variations in skin color. Nickell notes the manatee has no horn but suggests the witness saw one of the manatee's pectoral flippers juxtaposed beside its head when it rolled over in the water.

These appendages also have three vestigial nails which could explain the three-toe tracks. And manatees are known to "pec walk," partially leaving the water to graze on nearby plants, explaining the crushed brush and chewed upon weeds. Nickell also points out that manatees can thrive in fresh or salt-water, are found in rivers and the Gulf of Mexico, and have some notoriety for their travel range. He recalls a news story in 2006 about a manatee that traveled 720 miles up the Mississippi River to the Wolf River before succumbing to the cold at McKellar Lake outside of Memphis, Tennessee.

Regardless of the identity of their local monster, Newport, Arkansas, has embraced the aquatic mystery. Whitey has not been seen recently, but the Chamber of Commerce has not given up hope, sponsoring "Monster Nights: Newport's Downtown Entertainment Series." The series includes dining, concerts, night time activities, but no fishing tourneys.

## MS *Amerika*

In 1937, among the spike in Caribbean sightings, there was one that never made the media of the time. On October 26, 1937, the Danish East Asiatic Line's MS *Amerika* was approaching St Thomas. The ship had departed Southampton eight days earlier on a voyage to Vancouver via the Panama Canal, with stops in the Caribbean for the passengers and cargo stops along the West Coast. The witness was Dr. George Cooper of Budleigh Salterton, a small town on the coast in East Devon, England. Dr. Cooper was a retired radiologist and general practitioner, a career that had given him the habit of jotting down notes that resulted in a number of articles in the *British Medical Journal*. Even after his retirement, he kept a journal as he and his wife, Mary, began to travel.

News of Cooper's sighting didn't appear until the March 5, 1953, issue of the BBC radio guide *Listener*. His letter was in response to an article in the previous week's issue by naturalist John S. Colman on historical sea serpent sightings. Cooper wrote in because he thought his own account was more recent and might be of interest to Colman. Cooper was comfortable repeating his own story even though it was from 16 years earlier. As he noted, he kept a travel journal and wrote down the specifics of the serpent, which he then verified with the other witnesses.

The weather was clear and bright, when a quarter-mile away, heading in the opposite direction, "an elongated mass, sixty to eighty feet in length, broke surface in a flurry of foam and spray." The creature was chocolate brown in color, with a serpentine neck that thrust out of the water, followed by six or more protruding humps. A "flattish head" was angled as to give the beast and almost equine profile. It moved forward in a sinuous motion that made it look faster than it actually was; Cooper estimated the speed at 15-to-20 knots. Cooper made a particularly atypical observation—the creature was churning up a white wake that extended far behind the serpent.

Heuvelmans was perplexed by this sighting. The violent surfacing, serpentine neck, and pointed head seemed similar to a report he had from 1934 of a two-hump serpent breaching off the Azores, but the number of humps was the problem. His solution was that this creature was related to a pinniped, explaining that there were photographs of seals and sea lions in motion that showed folds of skin and layers of fat could form a series of waves (or humps) from surface friction when the pinniped swam at high speed. Heuvelmans has basically suggested the sighting on the *Amerika* is one of his theorized long-necked seals without actually mentioning long-necked seals. This could merely be an oversight, but it is interesting to note that this creature includes features that would qualify it for multiple categories in his classification system, creating issues with his system in the chapter of his book before he even introduces it.

## St. Andrew Bay Sea Cat

In March 1943, nature writer Thomas Helm was sailing with his wife, Dorothy, in a catboat in St. Andrew Bay. After some fishing and a brief thunderstorm, they were heading back to Panama City. At about 4 p.m., they were still a mile from shore, passing Redfish Point. The sea was so calm as to be almost slick, and the breeze was light.

Suddenly Helm saw something swimming toward his boat. The creature had a head the size of a basketball (a regulation basketball is between 29½ and 30 inches in circumference). The head rested on a long neck that reached four feet out of the water. More concerning was that the cryptid was only 30-to-40 feet off the starboard bow. A catboat—single-masted, 17-foot long, shallow draft, and barely moving along at two knots—was not a situation in which you wanted to encounter an unrecognized animal of undetermined temperament.

Helm adjusted their course to give a wide berth to the creature. As they sailed slowly around it, the couple was able to observe the beast as it turned its head and kept an equally close watch on the Helms. The head and neck were covered in wet fur that was "a rich chocolate brown." The eyes, set in the front of the head, were round, the size of silver dollars (1½ inch diameter), and glistening black. He saw what appeared to be a flattened nose and a mustache of stiff black hairs obscuring the mouth. There was no evidence of ears.

Helm's first thought was an otter or a long-necked seal, but he was familiar enough with seals, otters, and sea lions to realize this was not the correct facial structure. Aside from the lack of ears, the head most reminded Helm of a cat.

The creature looked away in apparent disinterest, then suddenly whipped its head back around and suddenly dived underwater. A mass of foam roiled up along the boat. The Helms did not see the beast again. Arriving at the dock, the couple spent several days confirming each other's account and then quietly making inquiries, first among the local marine science community, then the commercial fishermen. No one recognized the creature.

There was no doubt in Helm's mind it was not any pinniped he was familiar with, and he was well aware there are no pinnipeds native to the Gulf. He also considered the Caribbean monk seal (*Monachus tropicalis)*, which he mistakenly believed had become extinct two centuries before—the last confirmed sighting was actually in 1952, nine years after his encounter. More notably, Francis W. Taylor, president of the Warren Fish Co., of Pensacola, Florida, had written to naturalist Francis Harper in 1936 about Caribbean Monk Seals, and that on numerous occasions in the past fishing vessels had brought these seals into Pensacola, less than 75 miles from Helm's location.

Heuvelmans took notice of the case, pointing out that although Helm was correct that there were no pinnipeds with that long of a neck, he was incorrect in insisting that no pinniped had a head that large—both species of elephant seal (*Mirounga angustirostris* and *M. leonina*) would have a head in that size range. However, the long neck would preclude such an identification, and neither species is indigenous to the Gulf of Mexico.

Instead, Heuvelmans considered it as an example of his Merhorse class of sea serpents, in spite of having specified the criteria include a horse-like head, not the cat-like head such as Helm insisted he had seen. Heuvelmans skirts the issue by deciding Helm encountered a juvenile merhorse, which would have a more rounded head.

Although Helm claimed familiarity with various marine mammals, a furry basketball-sized head on a four-foot neck might allow a little leeway for some cognitive dissonance, especially if it's a rarely discussed cryptid such as a giant long-necked otter.

British Museum zoologist Maurice Burton published a book the year before Helm published his book with his sea cat encounter. Burton debunked most sightings of the Loch Ness monster as mats of algae and weeds, raised to the surface by methane as they decompose, with the occasional dead tree to give the illusion of the plesiosaur's neck and head. Burton, who also advocated giant eels as the source of sea serpents, admitted that not all of the sightings could be dismissed as decaying plants and misidentification of known animals. He proposes that the rare Nessie sightings left unexplained might be a long-necked otter, which could rise up four feet, compared to one foot for a standard otter. This would be a different, smaller cryptid than Heuvelmans suggested with his "super-otter" category of arctic-dwelling cryptids with pointed, seal-like heads and a 60-to-100 foot length.

Shuker, although willing to consider a warm-blooded, furred plesiosaur to explain sea serpents, was intrigued by reports of unknown long-necked giant otter, primarily in northwestern Ireland. Known as the *dobhar-chú* or "master otter," reports date back as far as 1684. As he notes in a chapter of his 2003 book, *The Beasts That Hide from Man,* being a creature of folklore does not preclude a cryptid from being a legitimate animal.

### The Camp Helen Creature

Powell Lake is the largest coastal dune lake in Florida. Such lakes are rare ecosystems—outside of the Florida Panhandle, these lakes are only found in Australia and New Zealand. The Florida version are freshwater bodies that have intermittent connections to the Gulf of Mexico. The inlet serves as a water level control, opening to allow water to enter or drain until it reaches equilibrium and then closes again. Phillips Inlet connects the Gulf and Powell Lake and is not maintained for vessel navigation. It meanders naturally from day to day, opening and closing as needed. It does, however, mean the lake has occasional visitors from the Gulf. Locals still recall Nuggie the porpoise who adopted the lake in 1965-66.

On Tuesday, January 27, 1959, Henry Gainous was fishing in Powell Lake when he spotted an animal carcass in shallow waters near Latimer's bayou, across the lake from Phillips Inlet. Gainous, a second-generation commercial fisherman, was stymied by the remains. It was obviously some sort of marine animal, four-feet long and weighing about 40 pounds. It had a dorsal fin and an oversized jaw with three rows of teeth similar to a shark's in size and shape. From dorsal fin forward, it had an elongated neck and a tough, hairless hide. From the dorsal fin back, it had porous skin and thin fur about three inches long.

Gainous took the creature to Emanuel "Bud" Adams, operator of a local fishing camp. Bud couldn't identify it either. The two remembered that a neighbor had claimed to have seen a lake monster on the west bank of the inlet, at Camp Helen, a vacation camp of small cabins arranged beside a large lodge dating back to the early 1930s. They brought the carcass to Howard Padgett, the caretaker at Camp Helen, to see if he thought it was the creature that had caused the panic the previous August. Padgett, who hadn't gotten a good look at the beast the previous year, decided it could be. They decided to call Captain Wally.

Captain Wallace Bradford Caswell, Jr., was a legend around the region. As a teen, he had started running a commercial fishing vessel but had grown so irate with sharks eating his catch and damaging his nets that he began leaping off the boat and jamming his arm in the shark's gills "until it drowned" (suffocated actually). He eventually realized a knife was more efficient. Caswell had parlayed this into a stunt show, visiting coastal towns to grapple with the local sharks. By the time he was 30 years old, he had killed more than 300 sharks and done underwater stunt work for Hollywood.

By 1959, Captain Wally looked more than his 54 years, with a silver plate in his skull from a swordfish battle and a bad leg from a sawfish tussle. Now running a whole fish market, a commercial fleet, and a detective agency, he was considered the local aquatic wildlife expert.

But Caswell had no formal training in marine biology, and the closest he had ever come to performing a necropsy was in his youth when he gutted sharks for entertainment. Caswell's examination seemed to focus on the assumption that the creature was a mammal, not a fish. He reported that the animal had nostrils, a ribcage, and lungs, but the lungs were so underdeveloped that the animal could only stay submerged for short periods. The jaw structure suggested a carnivorous beast; the teeth were shark-like. The flippers were rounded and would have only allowed limited mobility on land. There was no sign of ears or earholes, and the liver was unusually thin and long. Captain Wally also made sure the *News-Herald* in Panama City was aware of the monster.

Caswell decided the best thing was to cut off the head and ship it to the Smithsonian for identification. Caswell was correct that Smithsonian would help. Four days later, the carcass was identified as a decomposed blacktip shark (*Carcharhinus limbatus*). With the identity known, Caswell's description can be adjusted. Sharks do have nostrils, but unlike human nostrils, they are used solely for smelling and not for breathing. The liver and jaw are precisely what one would expect in a shark. The observed porous skin and thin fur were actually the skin decaying and allowing the underlying muscle fiber to break up into hair-like strands. The underdeveloped "lungs" were actually the shark's testes, located near the pectoral fins (the sperm ducts travel the length of the body). This was the end of the sea monster as far as the media was concerned. But was it?

The 1959 shark carcass did not match the description of the 1958 encounter. Howard Padgett, the caretaker at Camp Helen, had chosen to downplay the encounter the summer before, and the 1959 story in the *News-Herald* was the first time most locals had heard of the Camp Helen creature. Howard Padgett's reticence to promote the 1958 encounter was logical. Camp Helen's owners, Avondale Textile Mills of Sylacauga, Alabama, operated various mills and offered employees a week vacation each year at the coastal retreat, and Padgett suspected the owners would not appreciate their caretaker frightening off vacationing employees with talk of sea monsters.

The second reason was fear of ridicule; the front-page illustration of the "Creature of Powell Lake" proved his point. *News-Herald* cartoonist Wally Reichert rarely appeared on the front page and he made the best of it by disregarding biology. The drawing shows a mishmash of species—the head and long neck of an Apatosaurus, the dorsal fin of a shark, the body and fins of a manatee, and the tail of an alligator—all in one four-foot animal.

In the summer of August 1958, Howard Padgett had been teaching his 11-year-old son Eric how to water ski. Howard watched as Eric lost his balance and fell off the skis. It was then that Howard saw a large head protruding out of the water and heading toward Eric at high speed. Howard didn't recognize the creature and assumed his son was in danger. He swung the boat around between Eric and the creature, plucked the boy out of the water, and headed back to shore.

Only once on the boat did Eric finally see the creature, recalling it as "greasy black." The animal was still swimming with its head out of the water, occasional submerging, but not pursuing the boat. His father cleared the water of swimmers and went in pursuit with other boaters. Although they spotted the creature, it was too fast and maneuverable in the shallow water. Howard Padgett never discussed what he thought he encountered in the water that day.

Eric, however, retold the story as it became a part of local lore. After decades with the Panama City Beach Police Department followed by a return to Camp Helen State Park (the property was sold to the state of Florida in 1994) as assistant manager, Eric, now in poor health, was moving to Georgia. But before doing so, he agreed to visit Camp Helen in 2005 one last time to tell the story of his encounter, on Halloween, of course.

He also spoke to S. Brady Calhoun at the *News-Herald.* Padgett recalled the event clearly—after all, the creature had made him a minor celebrity. With 50 years of hindsight, he believed the beast that went after him was an otter, recalling that otters used to visit Camp Helen and attempt to scare off the picnickers long enough to grab a quick meal. His theory was that an otter was injured by a boat and the injuries left deformities.

Of course, Padgett had never seen the 1959 carcass, but he agreed with his father that the thing that washed ashore that day was the same creature, regardless of the Smithsonian's opinion. Calhoun also asked Michael Brim, a retired marine ecologist with the U.S. Fish and Wildlife Service, who carefully worded his reply. The Smithsonian, Brim suggested, could tell a shark from an otter when presented with a carcass. Brim also noted that Padgett claimed the animal popped its head out of the water, a behavior consistent with a river otter, not a blacktip shark. Brim suggested Occam's razor be used. Since Eric had never seen the carcass of the second creature (the blacktip shark) in 1959, a different animal may have chased him in 1958, one that could pop its head out of the water, such as an otter.

If that is the case, the 1959 description of the creature can be disregarded. The 1958 details can be gleaned from the various newspaper article recollections of the Padgetts. Those details are a creature swimming with its large, "greasy black" head protruding out of the water, occasional submerging. The creature was fast and maneuverable in the shallow water. It sounded and behaved like an otter, but Howard Padgett had been the camp director since 1946, replacing his father William H. C. Padgett, the first director at Camp Helen. Three generations of Padgetts were raised on the shore. No matter how otter-like the behavior was, if Howard Padgett didn't recognize the creature threatening his son, it was not an otter. In fact, the creature sounds similar to the "sea cat" spotted by Thomas Helms in 1943, barely 20 miles down the coast in St. Andrew Bay.

One possibility is the extinct Caribbean monk seal (*Monachus tropicalis*). The last confirmed sighting of a Caribbean monk seal was in 1952. These seals lived 20-to-30 years, so experts believe that some survivors survived into the 1960s or 1970s. Although subsequent searches to locate living specimens have failed, sporadic sightings continue to tantalize, including one that provides circumstantial evidence for Caribbean monk seal survival as late as 1997. Also enticing is a study by Adam and Garcia that includes photographs of Caribbean monk seals taken during a 1900 expedition to Arracifés Triangulos off the Yucatán peninsula showing a seal with dark fur and an oily sheen.

Some cryptids could be offered as alternatives but none that are typically associated with marine mammals in the Gulf of Mexico. Shuker's research on the Irish long-necked otter includes the 1722 gravestone of a woman reportedly killed by the *dobhar-chú*. The image on the stone suggests an otter-like head on a long neck, and a sleek body reminiscent of a greyhound with oversized paws. Of course, the problem would be explaining how an Irish long-necked otter ended up in Florida.

Eric Padgett passed away in Georgia on April 11, 2011. A memorial service was held in the park lodge at Camp Helen, where a small display on his encounter with the Camp Helen Creature remains on display. His legacy may turn out to be discovering an Irish cryptid had joined the ranks of snowbirds.

## The Lake Goose Waterhorse

In Michael Newton's 2007 book on Florida cryptids, *Florida's Unexpected Wildlife*, he begins his chapter on lake monsters by lamenting the fact that the major problem with the story of freshwater cryptids is that researchers are more focused on compiling lists of the monsters and too busy borrowing from previous author lists to bother verifying details about the lake or the sightings. He then proceeds to do the exact same thing. And the best example is Goose Lake. He paraphrases an October 25, 1881, article that appeared in the *Chester Daily Times* in Pennsylvania, the same item that was reprinted in Jerome Clark's *Unnatural Phenomena* two years earlier.

The article reports that fishing parties on Goose Lake were seeing a monster. James Z. Scott had seen it first, three years before, and reports had continued sporadically. Witnesses placed the body at 15-to-20 feet in length with a circumference as large as a horse. The head was dog-like yet had a tail like a catfish. No fins or feet were noted, but its movement was more fishlike than snakelike. All visible portions were covered with long, dark-colored hair. It also appeared to have an unsettling tendency to follow lighted boats at night.

The most recent encounter was on July 27, 1881. Aaron Terry and N. C. Osborne were out gigging from a boat in the evening. When the creature approached close enough, the gig was driven into the beast. The animal freed itself with a violent effort, twisting the prongs of the gig like so many straws. The article ends with a note that the locals were considering various options, including ordering harpoons and going on a whaling expedition. It sounds like a textbook example of Heuvelmans' long-neck seal, or under the revised Coleman/Huyghe classification system, a water-horse.

There are some issues with the newspaper story. There are three Goose Lakes in Florida, located in Pasco, Putnam, and Volusia Counties. Newton places the encounter at Goose Lake in Putnam County. How he came to that choice is irrelevant since the encounter actually took place at Goose Lake in Van Zandt County, Texas.

The story originated with the weekly *Wills Point Local Chronicle* and was picked up by larger Texan papers, which in turn, put it on the news wires. Somewhere along the way, someone mistook Wills Point for Willis Point, which is a Florida place name on the St. Johns River.

The Florida lake monster version ran on the East Coast (some versions cite the source as the *New York Sun*), including the Chester, Pennsylvania, article that modern writers cite, while the Texas version circulated in the west. In case there was some doubt, all the names in the article appear in the Federal Census as residents of Van Zandt County.

### Caribbean Monk Seals

The Caribbean or West Indian monk seal (*Monachus tropicalis*) was officially declared extinct in 2008. Of course, it was also declared extinct in 1983, but the 2008 designation was the procedural culmination of a process that involved multiple searches for the seal over the years until it was removed from the endangered species list. The seal had not been sighted since 1952, but hope had lingered on.

Adult seals were brown with a grayish tint on the back and yellowish-white on the belly and muzzle. They ranged from seven-to-eight feet in length and weighed about 350 pounds. These seals were similar in appearance to Hawaiian monk seals (*Monachus schauinslandi*) and Mediterranean monk seals (*Monachus monachus*), both of which are critically endangered.

Historically, the range of Caribbean monk seals was the Florida Keys, the Bahamas, the Greater and Lesser Antilles, and the Yucatan Peninsula. The last surviving colony was recorded in 1948 by C. Bernard Lewis, director of the Institute of Jamaica. It was located at Serranilla Bank, a partially submerged reef, with small uninhabited islets, halfway between Nicaragua and Jamaica. It was remote, but not remoted enough to escape their main predators—humans. A 1973 search for surviving seals discovered that the islets show evidence of fishing camps, suggesting the crews had slaughtered the seals, their perceived competitors for fish, although just the presence of humans on the islands was enough to drive the colony elsewhere.

Solow used math to calculate the probability of the Caribbean monk seals surviving into 1993 at less than one percent, a methodology that Boyd and Stanfield find fault with, in a 1998 article "Circumstantial Evidence for the

Presence of Monk Seals in the West Indies." Boyd and Stanfield note that Solow's equations are based on reports in scientific journals. They point out that the seal was already rare when John Edward Gray formally named and classified the seals in 1850. The decline in numbers continued from Gray's original assessment. By 1887, J. A. Allen of the American Museum of Natural History called it an "almost mythical species," and the scarcity of the animal was reflected in the paucity of research papers.

Instead, Boyd and Stanfield interviewed fishermen in northern Haiti and Jamaica in 1997 to assess the probability of the seals surviving. The fishermen were asked to select photographs of marine species known to them from randomly arranged pictures; 22.6% selected Caribbean monk seals. Delving deeper, the interviewers found 16 of the fishermen who had seen at least one seal in the previous two years. Circumstantial evidence suggested the Caribbean monk seals might not be extinct.

This is why the Caribbean monk seal continues to fascinate—the seal won't stay extinct. An unconfirmed report of one seal on an island off Haiti in 1985 was sufficient for the US Marine Mammal Commission to send out a team to investigate. The team found no evidence that the seals had been on the island but also concluded it was still too early to assume the species was extinct.

Mignucci-Giannoni and Odell disagree in their 2001 article in the *Bulletin of Marine Science*. They examined 20 sightings of unidentified pinnipeds in the Caribbean and Gulf of Mexico from the 1950s to 1995. (This list does not include sightings from cryptozoology circles, such as Helm's cryptid in St. Andrews Bay or the White River Monster.) They instead declare that the sightings are of hooded seals (*Cystophora cristata*) moving further south than their normal arctic range. Why the seals are traveling thousands of miles is not understood, and theories range from overfishing in the Arctic to overpopulation. But the phenomenon was first noticed in the northeast in the 1960s, which is when unsubstantiated Caribbean seal sightings began to increase in Florida and the Caribbean.

In 2014, a research team led by Dirk-Martin Scheel of Berlin's Leibniz Institute for Zoo and Wildlife Research looked at the genetic and physiological relationship between the three species of monk seals by examining the skeletal remains and skin samples in the museums that worked so hard to get examples of what they knew was a rare animal. His team noted that a disease might have delivered the final blow to the Caribbean Monk Seal in a losing battle that started in 1494 when Columbus mentioned his crew had killed eight seals. Humanity has been hunting the seals as a convenient source of oil, fur, and meat never since. The Caribbean monk seals, noted for their nonaggressive behavior, never stood a chance. The Caribbean or West Indian monk seal (*Monachus tropicalis*) was officially declared extinct in 2008.

Fortunately, in cryptozoological terms, extinction is a subjective term.

# CHAPTER 13
## The Legend of Three-Toes

The first sighting was an anonymous phone to the police on Thursday night, February 26, 1948. The caller claimed "some large sea creature" had badly frightened his girlfriend while they sat on Clearwater Beach, Florida. The police logged the call and did nothing. Two days later, early morning beachgoers spotted tracks coming out of the water just south of Everingham's Pavilion that continued along the beach for several hundred feet before returning to the sea. The tracks were of a three-toed foot, measuring 14 inches wide, 15 inches long, with the center toe longer than the outside two. Impressions at the end of each toe indicated toenails.

The *Clearwater Sun* covered the footprints in the February 29 issue but made it obvious what their opinion was on the whole matter, writing it off as a prank. Associated Press picked up the story and offered a few more details that the *Sun* chose to overlook to minimize panic. The *Miami Herald*, for instance, noted that the tracks came out of the water and crossed the beach to the seawall; there the tracks indicated the creature appeared to make four unsuccessful attempts to scale the 4½ foot wall before returning to the water.

That night, the three-toed phantom came ashore again, walked a few feet down the beach, then walked back into the surf. Snowbird Earl Hayes of Robinson, Illinois, collaborated the reports with his story of seeing something while fishing in the grass flats south of Big Pass—and the legend of Thee Toes was born. Admittedly, Hayes' "hideous looking something" sounds suspiciously like a manatee grazing in the grass flats, but any concerns were for naught. Traffic to the beach increased as the story was picked up by other newspapers across the US and Canada. The Clearwater paper chose to approach the growing interest as lightly as possible. A week after the tracks appeared, the *Sun* summarized the influx of theories with varying degrees of flippancy ranging from alcohol abuse to Dinny the Dinosaur from the *Alley Oop* comic strip (the latter was an inside joke, Alley Oop artist V.T. Hamlin was a Clearwater snowbird). One letter, mentioned in passing and then forgotten, was from Aleko Lilius, a Finnish explorer and journalist, who was interested in information on the tracks, which he thought resembled tracks he had photographed in 1937 on the shore of the Indian Ocean in what is now called the KwaZulu-Natal province of South Africa. Lilius was, at the time of his query, working for *Holiday* magazine in their Montreal office. If nothing else, the Lilius letter demonstrated that less than a month after the initial sighting, coverage of the "Clearwater Monster" had gone international.

Locals were quick to offer suggestions. One suggested a saltwater crocodile had traveled north up the coast beyond their normal range. Another suggested a sea turtle looking for a nesting place. The *Clearwater Sun* and the Chamber of Commerce were concerned that if it evolved into a full-blown "sea monster story," it would negatively impact the small town's tourism trade. The growing post-WWII tourism economy had all but been destroyed the previous year with a massive red tide bloom that lingered from November 1946 to August 1947. Tarpon Springs had been a ghost town—the red tide had decimated the sponge beds. Up and down the coast, "The spray of the poisoned surf inflamed human throats and lungs," according to *Time* magazine. "Tourists deserted the hotels; schools were closed; beach areas evacuated." The last thing Clearwater needed was a sea monster scaring off the vacationing guests.

While the Clearwater newspaper was trying to downplay the "monster tracks," the *St. Petersburg Times* was having a field day with the story. But when three-toed tracks appeared on March 19 in neighboring Indian Rocks, even the *Clearwater Sun* couldn't ignore the story. A county highway patrolman received a call that residents were in a near panic north of the Indian Rocks Inn when a sea monster appeared. The caller, John Moore, claimed it had chased him. The patrolman considered the tracks legitimate because of their depth in the hard sand. The paper still reminded people that in Clearwater proper, the two previous appearances of the tracks were considered a prank.

The St. Petersburg paper also believed the matter was a hoax. They pointed out that at 10:30 p.m. on the night of the 19th, a neighbor, Mrs. Woodrow Freeman, recalled that her dog starting yelping. This was just before John Moore called in the sighting. The trouble was, the newspaper reported, John Moore was a fake name. John Moore was the name of the local ghost that walked the beach since a fatal accident years before. The Freemans also saw a boat in the Gulf in the rear of their home, almost on the shore.

By the 21st, thousands of people had gone to see the tracks in the sand on Indian Rocks beach. And it was apparent that the police and newspapers had a fairly strong suspicion of who the prankster was. But without proof, no one was naming names. By March 26, the *Sun* was forced to admit that the tales were increasing as fast as the possible explanations, but the reporter also noted that as of that date, no one had actually seen the "monster." The latest explanation came from Jack Eckhart, a local resident. He claimed to have seen unusual three-toed sloth-like creatures in the salt marshes around Morro Castle in Havana harbor in Cuba. Eckhart recalled the Cuban creature walking on its hind legs, leaving three-toed prints. It had short arms with five claws and a head similar to a crocodile's but with a much shorter jaw. Eckhart saw it swimming outside the shark nets delineating the beach area, adding Cuban fishermen claimed these creatures could grow up to 30 feet tall!

Since there was no actual description of the Florida creature, Eckhart appears to be equating Three-Toes with an entirely different cryptid, one of the presumed extinct ground sloths. Although there is some debate in cryptozoology as to whether small pockets of ground sloths have survived into modern times on the Caribbean Islands (Ivan Sanderson believed there might be surviving ground sloths in the caves of Belize), the largest Cuban ground sloth (*Megalocnus rodens*) was four-to-five feet in height with no fossil record even hinting at the height Eckhart described.

The newspaper article, as did most theories, focused on the distance between the prints, which seemed to fluctuate from one account to another. Eckhart echoed the opinion that an eight-foot stride was beyond a prankster, and using stilts was impossible because of the weight, based on the depth of the prints. Other details appeared in the article. An appearance of the footprints near Everingham's Pavilion included the lifeguard stand being tipped over and residual evidence that something had rubbed up against the wood. As the tourists began asking questions, locals obliged by adding more appearances, building an impressive collection of footprint sightings that were, at best, unreliable. A fish market owner in Clearwater told his customers that hundreds of footprints had been seen on Dan's Island, south of Clearwater Beach. Tracks were also reported in Honeymoon Island north of Clearwater. A local marine biologist remained convinced it was a practical joke. The prints matched no known animal, and a turtle, for instance, would leave evidence of dragging its body as well as tracks.

But the crucial piece of information in the March 26 article was that law enforcement was going to be patrolling the beach at night to help relax the nervous residents. Other than an unsubstantiated report of the tracks on the shore near the Tocobaga ceremonial mound in Phillippe Park in Safe Harbor, on the opposite side of the Pinellas peninsula, there were no footprints sighted locally until October. There were, however, several reports that became attached to the growing legend of Three-Toes.

On July 23, two fliers from the Dunedin Flying School saw a creature a few feet offshore from Hog Island. The two aviators were George Orfanides, an out-of-work sponge diver and recent graduate of the school, and John Milner, owner of West Coast Pumps, a repair shop at the airport. The two returned to the airport to get more witnesses. Francis Whillock, owner of a beachside restaurant, and Mario Hernandez, the owner of the flying school, boarded the plane. They then passed over the creature five or six times at 500 feet and described it as "the shape of a log, 10 or 15 feet long and moves in a sidewise direction." It was covered in black fur with white legs or flippers and a hog-like head. The description varies between the Clearwater and St. Petersburg articles, perhaps more than might be expected by viewing the object from a moving airplane 500 feet in the air.

Generally, the descriptions sound similar to a manatee. Manatees will occasionally sport algae growths. They are so slow-moving that algae can take hold, and an algae growth on a manatee's back would give the appearance of brown fur. Milner, being a recent arrival from Chicago, could be forgiven for not recognizing a manatee, but Hernandez, Whillock, and Orfanides had all been residents of Florida long enough to have seen a manatee.

The footprints appeared again on Sunday, October 17, 1948, but the Clearwater Three-Toes was not in Clearwater. The tracks were discovered behind the Suwannee Gables motel on US 19, between Old Town and Fanning Springs along the Suwannee River. The tracks were 110 miles up the Gulf coast and 35 miles upriver and the first time that Three-Toes tracks had been found in a freshwater environment. Suwannee Gables owner Marx Cheney had gone out in the morning to inspect his "Jungle Drive," a path through the Spanish moss-draped swamp oaks along the river that Cheney used for horse-drawn carriage tours. With him was a guest at his hotel and his nephew, home on leave from the military. As in Clearwater, the tracks came out of the water, followed the bank, and then headed back into the water. The coverage by the *St. Petersburg Times* included a sidebar that would prove to be a pivotal moment—Ivan T. Sanderson was planning a trip to investigate the Clearwater monster case.

In 1948, Sanderson was becoming well-known as a naturalist and author. He had just begun appearing on WNBC doing a daily 15-minute radio program on nature topics and appearing on television talk shows displaying animals. While in Florida, Sanderson would record material on local flora and fauna, but there is no question Three-Toes was the focus of his visit. He had recently written an article for the *Saturday Evening Post* on the possibility of dinosaurs surviving in Africa, and here was evidence in the form of dinosaur tracks.

While scouting the Suwannee River from the air, Sanderson spotted a "dirty-yellow" creature roiling up the water. He estimated it was 12-feet long and four-feet wide, again within the normal range of a West Indian manatee (*Trichechus manatus*). Considering a small plane typically must maintain a minimum speed of at least 65 mph to stay aloft, and regulations mandated that planes maintain an altitude of 500 feet above the surface, spotting any-thing in the river was remarkable. The pilot was unable to circle back and relocate the spot where they had first seen it among the innumerable bends in the Suwannee.

In addition to the pilot and his sound engineer, Sanderson was accompa-nied by John O'Reilly, the more prominent reporter of the two, who reported nature and animal stories for the *New York Herald Tribune*. It was a high-pro-file expedition with O'Reilly and Sanderson reporting from Suwannee Gables. O'Reilly stayed a week and left convinced it was a hoax. On November 10, O'Reilly's article ran in the *New York Herald Tribune*. O'Reilly found that most of the locals already suspected it was a hoax. He was also fascinated by a Clearwater resident named Al Williams who flew up with a pilot friend to inspect the tracks. O'Reilly found out that Williams was a notorious practi-cal joker and the Clearwater Police Chief's most likely candidate for making the tracks. Williams had a habit of emphasizing how difficult it would be for humans to make the tracks. O'Reilly conferred with Dick Bothwell of *The St. Petersburg Times* who had been covering the tracks from the beginning. Bothwell, whose coverage of the Clearwater Three Toes was more joke than journalism, expressed his opinion that there were "public-spirited citizens in Florida who will go to considerable length to attract tourists" to their part of the state. O'Reilly's echoed the sentiment, noting that the only one who was taking the prints seriously was Ivan Sanderson, who was studying a chart he had made of the tracks.

*Miami Herald* columnist Stephen Trumbull concurred with O'Reilly. Trumbull was convinced Cheney was responsible, noting that Cheney, in ad-dition to the Suwannee Gables hotel, also owned the Suwannee Gables res-taurant, the Jungle Trail, a boat rental operation, and a curio shop, which had now had become the home of several casts of the tracks. Cheney claimed not to know how the prints got there, but he certainly wasn't going to let a market-ing opportunity pass.

The day after the O'Reilly article, another article ran in the *New York Herald Tribune*. It reported there was no question in Sanderson's mind that the tracks were man-made. Three-Toes was a hoax. This second *New York Herald Tribune* was not by Sanderson, but someone at the paper who had either read Sanderson's 53-page study of the tracks or listened to the pre-recorded episodes of his program for WNBC. The newspaper leaked it to the public before Sanderson's radio show aired, scooping Sanderson on his own story.

The episode that aired on November 15 reiterated the second *Herald Tribune* article. Sanderson's measurements of the tracks along the Suwannee showed a 31-inch stride, significantly shorter than the four-foot stride reported in Clearwater. Either Three-Toes' gait changed based on terrain, or the hoaxer was growing weary.

Sanderson knew it was a hoax, but with no crimes committed, it was just a good old fashioned practical joke. Sanderson knew who it was—his correspondence files include letters between him and Al Williams, who had always been the prime suspect.

Now that the cat was out of the bag, several residents admitted they knew someone from Clearwater had walked along the shore with metal tracks strapped to their shoes.

On his return to New York, Sanderson found stacks of correspondence waiting, including from eyewitnesses who had also seen the "monster." Sanderson was discovering that his dabbling in Fortean zoology was becoming more popular than his wildlife writings. In spite of having debunked the Three-Toes, Sanderson kept these reports of sightings, most of which could be easily identified. A Milwaukee couple rented a rowboat and explored an island north of Tarpon Springs. Something waddled down the beach and slid into the water. The couple wrote to Sanderson that the creature was large and gray, and had "a head like a rhinoceros but with no neck." This description is closer to that of a vagrant pinniped such as a South American sea lion (*Otaria flavescens*) or a southern elephant seal (*Mirounga leonina*) than to the bipedal dinosaur that Sanderson was initially investigating, but others were even more obvious.

A series of August sightings came in from Chiefland, Florida, a few miles downriver from Suwannee Gables. A deacon of the local Baptist church was having a picnic on the riverbank when they saw something dome-shaped with a rough and knobby surface moving upstream. By the time they reached a motorboat to investigate, the creature had vanished. Three days later, a lady angler claimed to have seen something large and dun-colored paddling upstream as she fished on the bank beneath the bridge carrying US 19 across the river. The color matches the "dirty yellow" that Sanderson had spotted in the air. "Dome-shaped" is a picture-perfect fit for the back of a manatee, which would logical, considering that Florida's Manatee State Park is less than 10 miles downriver from these sightings.

In 1951, Sanderson would revisit the Clearwater Three-Toe story in a *True* magazine article. Ironically, there is no indication he had ever actually visited Clearwater itself, and there was no need—the tracks he examined were in Old Town. In spite of his declaration on radio and print that the entire matter was a spectacularly successful joke, the *True* article finds Sanderson doing an about-face, claiming the tracks were legitimate and possibly those of a bipedal dinosaur, but more likely a giant penguin. This realization made Sanderson reevaluate his reports of sea monsters. He found numerous reports in the Antarctic of a short-necked sea monster with staring eyes and a head shaped like a camel. Add that to the hump that appears when the creature swam along the surface, and he began to suggest that there was possibly an undiscovered colony of giant penguins on an island in the Antarctic Ocean.

Meanwhile, Al Williams and his partner Tony Signorini would occasionally put on the shoes and make new tracks, but it was an open secret among Clearwater residents. The only ones being fooled were newcomers, tourists, and the occasional cryptozoologist. And even individual tourists were in on the joke. In 1953, Arthur C. Clarke was in Clearwater as a guest of local physician George Grisinger. Dr. Grisinger was a great admirer of Clarke's recent book *The Exploration of Space* and an underwater photographer of local renown. He invited Clarke to try a new technology he had adopted, called the "aqualung." Clarke became an avid diver immediately. He wrote a letter to fellow British science fiction author Eric Frank Russell about scuba diving, and added a story about Clearwater's "enormous bird," knowing Russell was also an avid Fortean. "I was taken into a back room and shown the footprints, neatly built round a pair of boots. Whenever the character who owns them feels like a bit of fun, he puts them on and walks backwards down into the sea."

Others were less fortunate. When Ruth Dyckman, a novice reporter, came to the *Clearwater Sun* in 1952, Williams and Signorini waited until the high school graduation, when the *Sun* had the new reporter on duty late into the night, covering the various social events of the new graduates. An anonymous call at 10 p.m. reported the tracks and gave specific directions on where to find them. The reporter and a photographer rushed over to the Courtney Campbell Causeway that carried State Road 60 across Old Tampa Bay, connecting Clearwater to Tampa. The causeway is protected by riprap for most of the route but also has open sandy beaches, popular as fishing grounds and picnic areas. Exactly as described by the anonymous caller, the tracks were a mile out on a picnic beach along the causeway. As usual, the tracks came out of the water, walked along the beach, and returned to the sea. Dyckman found a Tampa family crabbing and fishing 500 feet away. They had been at the location since 7:30 but had not heard or seen anything. By coincidence, both Williams and Signorini had family living near the causeway.

By 1967, Sanderson had discarded any lingering doubts about dinosaurs or hoaxes and fully embraced his giant penguin theory, which first appeared in *Fate* magazine. His candidate was a bird that didn't even exist in fossil records, a 15-foot-tall, thick-billed penguin. In his later book *Investigating the Unexplained*, he identifies the New Zealand fossil of a narrow-flippered penguin (*Palaeeudyptes antarcticus*) as being seven-feet tall. Apparently, by Sanderson's logic, if a penguin could reach seven feet, it could reach 15 feet. In addition to the flaws in that reasoning, *Palaeeudyptes antarcticus* only stood between 3 ½ and 4 ½ feet tall. (The tallest penguin found in the fossil record, the six-foot, eight-inch *Palaeeudyptes klekowskii*, was not discovered until 1980 and is still far below Sanderson's 15-foot bird.)

Sanderson partially based his theory on the story told by the couple from Milwaukee in 1948 who described their Tarpon Spring creature encounter as waddling down the beach and sliding into the water, behaviors Sanderson equated with penguins in general. The two-part article in *Fate* was met with confusion and derision. Undeterred, his 1972 book placed the Florida Three-Toe squarely in the middle of a chapter on giant penguins and giant eggs as an international phenomenon. The section is jarring, with Sanderson claiming that many people had seen a 15-foot penguin in Florida, when the descriptions by witnesses were decidedly not penguinlike.

Al Williams had died in 1970, so there was no obvious candidate to counter Sanderson's increasing odd recollection of the incident. Tony Signorini, William's partner in the fakery, had taken over William's automotive repair company and was overwhelmed. He put the print-making shoes in a box, tucked it under a work counter, and forgot about it, even as the giant penguin discussion continued in cryptozoological circles. Sanderson died in 1973 having transformed the Clearwater prank into an example of a 15-foot cryptid penguin. No one is exactly sure what Sanderson thought personally, as opposed to his printed statements.

The mystery of the Clearwater tracks should have been finally put to rest in 1988 when, at the urging of friends, Tony Signorini contacted Jan Kirby, a freelance reporter for the *St. Petersburg Times* who covered Clearwater, and came clean on the joke. Al Williams was the mastermind, notorious for the variety and scope of his practical jokes. The two had been brainstorming a prank that would feed off the popularity of the Loch Ness. When dinosaur tracks appeared in the news in 1947, Signorini came up with the idea of making "dinosaur tracks" on the beaches. Early attempts with plaster and cement molded prints failed. The two then went to nearby Largo, where Tony Signorini's father, Angelo, had retired in Florida after a career as a steel moulder in Pennsylvania. He still dabbled in metalwork as a hobby, having built a small foundry in his yard. Tony knew enough of the process that they were able to create the tracks in cast iron. Holes were drilled to install rods, and a pair of high-top canvas sneakers were bolted to the tracks. The inner soles were glued back into place, and the "dinosaur" was ready to visit Clearwater Beach.

Victor Toro, a footwear designer, notes that the 1940s high-top sneaker was an excellent choice to withstand the strain of a 30-pound weight attached to it. The sneaker's construction was heated to bond the canvas to the vulcanized rubber soles. The rubber sole (which included a molded toe cap as opposed to modern toe caps that are separate pieces glued in place) added strength and support. Toro feels that the high top was one of the few readily available choices that could be used repeatedly in the prank without ripping apart.

Signorini retrieved the dinosaur shoes for Kirby and recounted his exploits as a sea monster. The four witnesses in the airplane who had seen the creature was a total surprise to Williams and Signorini. They were not in on the joke. But three months of lucrative curiosity seekers looking for monsters might suggest that the businessmen were hoping another sighting might coax the sightseers back again.

The Suwannee River tracks were thanks to a friend with a fishing lodge on the river. They had scouted several spots while staying at the lodge and left tracks scattered along the river. But it wasn't until Marx Cheney got involved that the tracks received national attention. Unbeknownst to the jokers, Cheney was a retired show business agent whose previous hotel in Noel, Missouri, had been used as the base of operations for the 1938 filming of the Tyrone Power film *Jesse James*. Cheney had purchased acreage that became the Suwannee Gables operations from the proceeds of that film, and he still had media connections. If Williams and Signorini had ever considered taking their little joke internationally, they couldn't have picked a better spot to leave tracks then in the backyard of man who knew Hollywood publicists and press agents.

Signorini notes how he and Williams carefully picked the nights the dinosaur would come ashore, based on wave action, darkness, and tide. Sanderson was unable to reproduce the prints to his satisfaction, going so far as to build a pair of monster-print shoes weighing 35 pounds per foot. Sanderson was likely unfamiliar with Florida soils; in trying to recreate the prints, he doesn't mention the extent to which he reproduced the soil conditions. On the night of October 16, when Williams and Signorini went to make the tracks along the Suwannee, there had been a typical Florida rainstorm—a moderate to heavy downpour at 6 p.m. that dumped .25 inch of water in less than an hour. Not only did that chase away any visitors, but it also allowed the shoes to sink easily into the ground that was already saturated from a hurricane two weeks before. By morning, when the tracks were spotted, the ground was dry at the surface. By the time Sanderson arrived on November 4, any attempt to recreate the tracks were on dry, hard-packed sand. The water had continued to leach back into the watershed, creating entirely different soil conditions compared to the original track making conditions.

Signorini's disclosures sent ripples through cryptozoology. Since the original *New York Herald Tribune* article with Sanderson's declaration that the tracks were a hoax was all but forgotten, most of organized cryptozoology and Fortean studies aficionados only knew the giant penguin theory.

Soon after the Signorini confession ran in the *St. Petersburg Times,* the Society for the Investigation of the Unexplained, the group Sanderson founded, released their newsletter summarizing the article. They admitted Sanderson was convinced it was a practical joke, but that "Ivan the investigator could also masterfully play the role of entrepreneur of mysteries," reviving the story for the sake of a good story in a paying market.

The International Society of Cryptozoology ran an abridged version of the Kirby article in their *ISC Newsletter,* effectively admitting the story was a hoax and blaming Sanderson for its longevity. The piece closes with a quote from James Randi, noted magician and vocal skeptic of the paranormal. In part, it reads that Sanderson "was in the business of writing books about strange subjects, and he would never allow ugly facts to interfere with an otherwise attractive story. In person, he left no question about his doubts; in print he successfully resisted expressing any really serious reservations he had."

Both articles echo a remark that Tiffany Thayer had made in the Fortean Society magazine *Doubt* in April 1949 in the wake of Sanderson's trip. Noting how Charles Fort would argue both sides of a debate equally well and never commit to a resolution, Thayer said about Sanderson: "He is variously quoted by interviewers. The tracks are a hoax—the tracks could not be a hoax. In fact: he out-Forts Fort in suspending his judgment."

Although the major cryptozoology and Fortean societies accepted that Sanderson had been fooled and moved on with their investigation, there remained a dedicated few who then proceeded to attempt to debunk the claims of a hoax, seemingly unaware Sanderson had all but single-handed kept Three-Toes in the public eye while knowing it had been a practical joke.

The most ardent defender of Three-Toes is Michael Newton, a prolific fiction writer who also specialized in encyclopedias on crime topics and dabbled in cryptozoology. He argued that Signorini's 40-year-old memories didn't match with Sanderson's documentation, which included secondary information, such as the number of times tracks appeared. Sanderson had based the size of his penguin on the required weight needed to replicate the depth, but soil conditions had drastically changed along the Suwannee River, and the tracks were over two weeks old by the time he arrived. Newton then recounts the various eyewitness who could not possibly mistake Tony Signorini for a 15-foot beast resembling a bipedal hairy rhinoceros, one of the various conflicting descriptions. His major issue is how easily the mainstream media just accepted the story as a hoax, instead of researching the story to prove Newton was right.

Author Greg Jenkins reiterates that there were eyewitnesses. Tony Signorini was obviously trying to take credit for a hoax that wasn't fake, just to "be known as the town funny guy." His argument of no skeletal remains and no artifact evidence is the traditional aphorism "absence of evidence is not evidence of absence" and outweighs the evidence of the hoax. The cast iron tracks, he inexplicably declares, were "hastily made."

Despite such defenses, the matter remains resolved. *Fortean Times* revisited the hoax in 1992, squarely blaming Sanderson for disregarding contradictory details provided by the eyewitnesses, including himself, which led to stories of the "Florida Giant Penguin" appearing in cryptozoology books. They wondered out loud if the different creatures were indicative of one of Charles Fort's "flaps," where scrutiny of one cryptid revealed a series of unrelated creatures overlapping the main interest.

In 1998, to mark the 50th anniversary of the Clearwater Three-Toes, the *St. Petersburg Times* sent a reporter to interview 78-year-old Tony Signorini again. Signorini recounted several of his and Al's other pranks. He recollected having strapped on the shoes around 20 times over the course of a decade. Al Williams was crafty. He knew the sheriff was well aware of who was behind the tracks but couldn't prove anything. Williams and Signorini serviced the patrol cars, so they knew the shifts and schedules better than some of the deputies, and when the department was short a patrol car (since it was in their shop). Williams calculated foot traffic, tides, and waves. When a call from Al came in at 2 a.m., Signorini often didn't know if it was to repair a broken-down squad car or to visit the beach with the shoes. It worked spectacularly well for a decade; the closest they came to being caught was when their boat was once spotted along Indian Rocks beach.

In 2002, Eberhart released *Mysterious Creatures*, which includes the Huillia, a Trinidadian cryptid. He notes it's the common name of the anaconda (*Eunectes murinus*) and then proceeds to describe something that is partly serpentine and partly nonsense, such as leaving three-toed tracks and having a high-pitched whistle. His sources are Joseph's *History of Trinidad* and a letter to the editor in *Strange Magazine*.

The Huillia's three-toed track reference began appearing as supporting evidence for the theoretical range of the Clearwater Three-Toes, along with the 1943 *Clearwater Sun* letter by Eckhart about tracks in Cuba. Eberhart's citations do not hold up to closer scrutiny. In Joseph's book, he related how he assisted in killing a 22-foot long "boa constrictor," a length that is both an exaggeration and a misidentification, according to Hans Boos, author of the definitive work on the snakes of Trinidad and Tobago. He notes that Joseph consistently confused anaconda and boa constrictors, and he was not the only author of that time to do so. Joseph's book never uses the word anaconda at all. More notably, the word Huillia never appears in the book, and neither do references to whistling or three-toed tracks. Day's 1852 history is the first to use the local name, using the variation Huilla, which is also the spelling Eberhart uses. It would appear Eberhart confused which Trinidadian history he used, but aside from the exaggerated size, there is little to suggest a cryptid.

Eberhart's second source is a letter to the editor of *Strange Magazine* by John O. Brathwaite of Trinidad. Brathwaite was seeking help identifying "a monster that has been described as a cross between a horse and a dinosaur. It has been sighted in a swamp by more than one person and has been seen both in the daylight and at night. Its color has been said to be moss green and [it] swims very fast. A high pitched whistle has been heard around the time of the sightings. It has a three-toed footprint and it appears as though there is more than one animal."

Without a confirmed sighting, the creature sounds like one of the monitor lizards, such as the Asian water monitor (*Varanus salvator*), whose tracks can be mistaken for three-toed. Much like the Florida environment that allows abandoned and escaped iguanas and monitor lizards to survive, a lizard could easily survive in Trinidad. Regardless of the creature's identity, it has nothing to connect it to the anaconda except geography. The Eberhart listing merely confuses matters with a hybrid of unrelated reports.

*St. Petersburg Times* revisited Three-Toes one last time, six years later. The reporter asked if Signorini wanted to go to the beach and make tracks. The 85-year-old was increasing frail but maintained his sense of humor. Unable to even carry the shoes, Signorini dutifully went to Clearwater's Sand Key. But Signorini could barely stand in the shoes.

It would turn out to be the last time that Tony Signorini would walk the beaches as Clearwater Three-Toe. He died in June 2012 at the age of 91. His obituary proudly lists his military record, his service to St. Cecelia Catholic Church, and his fame as the Clearwater Monster. His son Jeff inherited the family auto shop, along with a pair of black high tops bolted to 30-pound cast iron three-toed treads. Competing historical societies asked for the shoes for their collections as historical artifacts of a hoax that brought Clearwater international attention.

Jeff Signorini decided to keep the shoes in the family for the time being but did put them back to use one last time. In 2014, the cable television series *Mysteries at the Museum* used the shoes to make tracks on a local beach in a re-creation of the 1948 events.

In January 2016, the International Cryptozoology Museum in Portland, Maine, sponsored the first annual International Cryptozoology Conference in St. Augustine, Florida. Author Robert Robinson was in attendance and made a special presentation to Loren Coleman. Robinson had visited Jeff Signorini to discuss his father's legendary footprints. As part of the visit, Robinson was allowed to make a plaster cast of a freshly made three-toe track. That plaster cast was presented to Coleman at the conference for inclusion in the museum collection. Since the museum exhibits include both cryptids and known hoaxes, no matter which side of the debate you subscribe to, the Three-Toes footprint is a natural addition to the museum.

# CHAPTER 14
## Early Encounters

To early European explorers, the New World was filled with cryptids. Even the gentle manatee was unrecognizable—*Nova Typis Transacta Navigatio* by Caspar Plautius in 1621 includes an engraving showing a manatee more piscine than mammalian, and big enough for five Natives to ride on with room to spare. It doesn't help that the illustration is based on an anecdote from Francisco López de Gómara's book on the Hernán Cortés military campaign that was so inaccurate, Prince Philip (afterward Philip II of Spain) issued a decree banning the volume. Neither Plautius nor López de Gómara actually visited the New World.

Similarly, a 1784 book by Antonio de Ulloa insisted that the fearsome "blanket fish" would attack swimmers, wrapping them in their large fins, and then sink to the bottom to drown their victim. The book was translated into French and English, giving Europeans a wide variety of languages in which to fear the blanket fish, a literal translation of the Spanish term still used— "manta." The manta ray (*Manta birostris*) does indeed appear similar to the Spanish cloak of the time with a similar name, but the manta ray is only a danger to plankton and small fish.

### Giant Turtles
Giant turtle sightings are rare but global. Both Heuvelmans and Coleman/ Huyghe treat them as a separate category in their classification systems. Heuvelmans names the group "The Father-of-all-the-turtles" after the Sumatran folk figure, although he admits he's not convinced there is a giant turtle species since most of the sightings are in waters too cold for reptiles. Instead, he wonders if these turtle reports are merely creatures that belong in other categories (such as long-necks and merhorses) but were described as turtle-like by untrained observers.

The Coleman/Huyghe classification system calls them "Cryptid Chelonians" and agrees with Heuvelmans that they are rare and share characteristics, including a very wide mouth, prominent eyes, and large scales or a carapace on the back. Coleman specifically suggests a 30-foot length. Although both their books include examples, neither includes what is probably the highest profile figure among giant turtle witnesses—Christopher Columbus.

In early September 1494, Christopher Columbus' ships were sailing east along the southern coast of Hispaniola (Dominican Republic). As chronicled by Las Casas, on the approach to the island of Saona, they saw a "remarkable fish" as big as a medium-sized whale. The beast had a turtle-like shell described as looking like an adarga, the leather shield of the North-African Berbers, which has the distinctive shape of a pair of overlapping ovals. The head was slightly smaller than the barrel known as a pipe (1⅓ diameter and 3½ feet long), and it had a long tail with large "wings" on either side.

Columbus seemed unperturbed by the sighting, only noting that when such creatures surface, bad weather is coming. Soon after, a gale struck, but the little fleet, forewarned, was safely sheltered behind the island. His biographers are equally unimpressed. Samuel Eliot Morison, in his biography of Columbus, defers to Armas's 1888 book listing the animals encountered by Columbus. Armas professes some confusion over the description before deciding it was a short-finned pilot whale (*Globicephala melaena*). Neither Armas nor Morison explains why ships full of Spanish sailors suddenly did not recognize a frequent visitor to Spanish waters.

Environmental scientist Robert France offers a new theory in marine cryptozoology on sea monster identification, focusing on sea turtles. France suggests the long tail, rather than being a "real tail," could represent an early example of a turtle entangled in Native fishing gear such as fiber netting, creating the tail as the debris trails behind the creature. France goes further, suggesting this theory that sea turtles entangled in fishing equipment are responsible for some of the most famous sea serpent sightings with their serpentine movement being merely ropes and nets, and reports of humps being old net floats. Therefore, the decline in sea serpent sightings is a result of the introduction of more biodegradable artificial nets and ropes.

There does seem to be merit to France's theory. In 1921, the steamship *Munamar* arrived in New York from Cuba with a tale of an encounter with a sea serpent—of a sort. The ship observed a sea serpent merrily bobbing on the surface, coiling and uncoiling, much to the delight of the passengers. It was only when the captain brought the ship in for a closer look that the monster was identified as a barrel tangled in rope and trailing 30 feet of hawser line. This mistaken identity took place on smooth seas in bright daylight; it is not a stretch to imagine a turtle trailing netting being similarly misidentified.

German cryptozoologist Ulrich Magin suggests there are ample examples of turtles large enough to either be a new species or outliers on the size range of turtles (and large enough to survive pulling fishing tackle). Magin notes Gabriel García Márquez, before his Nobel Prize days, was a reporter for *El Espectador* in Bogota, Colombia. In 1955, eight crew members of the ARC *Caldas*, a Colombian destroyer, were washed overboard while returning to Cartagena from drydock in Mobile, Alabama. Ten days later, a lone survivor washed ashore in a lifeboat. Márquez interviewed the survivor, Luis Alejandro Velasco, for a 14-part newspaper article on Velasco's ordeal. The Colombia regime, it turned out, didn't like Velasco admitting it was not a storm that threw the crew overboard—the ship was so overloaded with American contraband appliance for the Colombian elite that the *Caldas* couldn't handle rough seas, nor could she turn hard enough to turn around to rescue the men. The interviews by Márquez were compiled into a book and released in 1970, the first appearance of some of the material (an English edition was published in 1986).

Of interest to cryptozoology is Velasco's ninth day of fighting sharks, starvation, and dehydration. As darkness set in, Velasco saw a giant yellow turtle with tiger stripes only 16 feet from his tiny raft. Unsure if it was a hallucination, he shifted in the raft. Seeing the movement, the turtle sank, leaving a trail of foam. Still not sure if it had been a hallucination, Velasco was certain that even a light brush from the beast would be enough to upset the raft and send him to the waiting sharks.

The account refers to the encounter as a "giant loggerhead," but that could be artistic license by Márquez. Although loggerheads (*Caretta caretta*) frequent those waters off Colombia, the Sea Turtle Conservancy notes only one sea turtle with a carapace color that includes brilliant yellow with radiating stripes in green and brown, the green sea turtle (*Chelonia mydas*). The green turtle may be the largest of the Cheloniidae family, but the largest specimen recorded was five feet, a far cry from Velasco's 13-foot turtle.

Magin notes other accounts of giant sea turtles observed in Newfoundland, Nova Scotia, Scotland, France, and Spain. He interprets this as all being the same species, connected by the route of the Gulf Stream. He finds that both nine-foot reports could be oversized examples of known turtles, but anything 14 feet or over (he calculates Columbus's "medium-sized whale" as 60 feet) must be a new species of marine animal.

Part of the problem in identifying Magin's theoretical new species is finding examples. A basic search of the internet for turtles using the adjectives "massive," "huge," or "giant" will bring a disturbingly large number of beached turtle carcasses, all well within the average size range of an adult turtle.

## De Orbe Novo
In 2003, two years after Bernard Heuvelmans's death, an English edition of his revised sea serpent book was released as *The Kraken and the Colossal Octopus: In the Wake of Sea-monsters*. In the aftermath of the *Biological Bulletin* debunking of the St. Augustine Octopus, Heuvelmans had included revisions and addendums for the English release, influenced by the work of French cryptozoologist Michel Raynal.

Raynal came across a reference to a monster encounter in the work of Pietro Martire d'Anghiera (1457-1526), suggesting that this was the first published account of a giant octopus attack in the Caribbean. In *De Orbe Novo*, as Raynal interprets a 1907 translation, d'Anghiera describes a monster coming from the sea and capturing a sleeping sailor about his middle and dragging him into the sea. Unfortunately, the myriad of translations of *De Orbe Novo* vary on key points. Going back to the original 1530 publication in Latin, the encounter in question took place at the mouth of the Orinoco River, on the Paria Peninsula in present-day Venezuela.

A more accurate translation would be that something came out of the water, grabbed a sailor sleeping on the sand, and jumped back into the water with the man shrieking. Considering the location, the explorers undoubtedly encountered the Orinoco crocodile (*Crocodylus intermedius*), rare and endangered today, but common at that time in the freshwater rivers of northern South America, particularly the river that gives it its name. It is a very large species, one of the largest predators in the Americas. Before the species' decimation for its skin, males were reported up to 22 feet. Perhaps not as spectacular as a giant octopus or squid, it was nonetheless just as lethal.

This lethal encounter was not an isolated incident. Sir Walter Raleigh and his men had a similar encounter on the Orinoco. Theodor de Bry issued a series of books on voyages to America between 1590 and 1634, including part 8 in 1599, where Raleigh describes the alligators and mentions a young Native eaten by alligators when he jumped into the river to swim. De Bry never visited the New World, and his illustration of the incident shows a sea serpent, not an alligator—it was the closest he could envision.

## The Marrajo of Las Casas
Even sharks were unfamiliar terrors. As shark scientist José Castro notes, large sharks are absent from medieval Europe's bestiaries. European fishermen fished

in rivers and along the coast, so encounters with large sharks were at best few and far between.

So when explorers first encountered large sharks in the Americas, they had no terms to differentiate these massive sea monsters from the innocuous small sharks with which Europeans were familiar. The Spanish borrowed the word *tiburón* from the Carib Indians, and the English adopted the name from the Spanish. The origin of the word "shark" is less definite, but appears to have been derived from the Mayan dialect Yucatec word *xoc* (pronounced like "shock"). It appears suddenly in English in 1569, which coincides with the return of the few surviving sailors of an ill-fated John Hawkins expedition that culminated in a pitched battle with Spanish ships in San Juan de Ulua, Vera Cruz. Tom Jones, a professor of European cultural history, hypothesizes that the ensuing bloody carnage was accompanied by a shark feeding frenzy that made the xoc a memorable word to the survivors.

And then there is the *marrajo*. Today, it refers to mako sharks, but in the 16th century, Bartolomé Las Casas, in his *Historia de las Indias,* used the term to specify a fish larger than the *tiburón*, one much larger and capable of swallowing a man whole—a great white shark (*Carcharodon carcharias*). This is why La Casas has the dubious honor of recording the first report of a shark attack on a human in the New World. A Spanish slave driver forced a pearl diver back into the water after the Native fled the water at the sight of a marrajo. The marrajo then attacked the diver. The other Natives were able to capture the shark with a baited hook, but by the time they landed the monster and cut him open, the swallowed diver gasped for air and died.

The great white was far more common in the coastal waters of the Caribbean because the beaches were still home to large colonies of the Caribbean monk seal (*monachus tropicalis*). When the seal colonies were decimated, the great whites moved on to other prey. Unfortunately for the native population, they fared no better under the Spanish than the seals did.

## Maps and Monsters

On the cover of this book is a sea monster in the Gulf of Mexico, reproduced from the map officially called *Floridae Americae Provinciae Recens & exactissima descriptio Auctore Iacobo le Moyne cui cognomen de Morgues, Qui Laudonnierum.* For obvious reasons, it usually referred to simply as the "Le Moyne Map," after French artist Jacques Le Moyne. The map is considered one of the most important 16th-century maps of the region. Le Moyne accompanied the French expedition of Jean Ribault and René Laudonnière in an ill-fated attempt to establish a French presence in North Florida by founding Fort Caroline on the St. Johns River (near present-day Jacksonville) in 1564.

After Le Moyne's death, the map was first published in De Bry's *Brevis Narratio Eorum Quae in Florida Americae Provincia*. The map's importance was that it contained new information on the coast and waterways of the Florida region. It is not entirely accurate (Le Moyne extends Florida too far easterly rather than northeasterly), but once it appeared in De Bry's book, it became the primary resource for other cartographers to "borrow" from for the next 150 years.

The double spouting sea monster is a common motif in maps. Cartography historian Chet Van Duzer attributes this to the influence of Olaus Magnus and his monumental work on the history of the Nordic lands and peoples, which includes a wide variety of sea monsters. This influence extends beyond cartography into cryptozoology—both Oudemans and Heuvelmans devote pages to Olaus's creatures.

Van Duzer notes that the images were widely used and modified, but the dual spouting baleena can be traced back to specific sources. If the beast is spouting forward, it was from Olaus's *Carta Marina* map. Backward spouting indicates the inspiration was book 21, chapter 9 of his *Historia de Gentibus septentrionalibus*.

Van Duzer further suggests that the purpose of sea monsters on early maps depends on the date. On earlier maps, the image could be a warning for a dangerous location or a mnemonic for sea monster literature, the cartographic equivalent of the exempla circulating among contemporary clergy who used the anecdotes to augment sermon. Jonah and St. Brendan are common examples. Later maps, the theory continues, would use these monsters as decorative elements to fill areas of the map not yet explored and reflect the Renaissance's increasing interest in wonders and marvels. Whether sea monsters on maps served in a mimetic or mnemonic capacity remains a point of debate.

# CHAPTER 15
## Folklore and Fakelore

Michel Meurger states in his book on lake monsters that cryptozoology's reliance on folklore and traditional stories is at best problematic. This "scientification of folklore," as he calls the practice, uses local tales as a record of an unbroken chain of evidence that adds credence to contemporary cryptids. The problem with this, he continues, is that from the Renaissance forward, naturalists were trying to rationalize evidence through the veil of a strong folklore tradition. This creates a fallacy that Heuvelmans is guilty of. In a 1982 article in *Cryptozoology*, Heuvelmans claims it is easy to detect a traditional story from a testimony. This would be true, Meurger notes, if folklore was static, rather than continually evolving and adapting to the times. A helicopter landing on the back of a cryptid in Lake Pohénégamook in Canada is just a modern variation of St. Brendan landing his vessel on the back of a whale.

Even with this cautionary note in mind, folklore is still considered a potential indicator of a theoretical cryptid. Whether such a cryptid is simply being compared to a folklore creature, or if the folklore actually documents the cryptid's history, remains the issue, but this mutability of folklore has a new ally—social media. Hence, the rise of what I refer to by the imprecise term "fakelore," the misidentification of something as a possible cryptid, as opposed to a deliberate hoax. Such reports were briefly considered cryptid but were discovered to have a more mundane origin. Unfortunately, an appearance on the internet injects a degree of permanency that means fakelore will show up, sooner or later, in a book as folklore.

An excellent example of an "almost a cryptid" fakelore is the encounter at Miami Beach on July 28, 1949. Seventeen-year-old Carol Krieg was swimming in front of the Cromwell Hotel with her parents when she saw the surf breaking over something "big and black and slimy." She immediately screamed that there was a sea monster. Her parents agreed and ran back into the hotel where a bellhop heard the commotion and investigated. He saw the sea monster and called the police. Two patrolmen were dispatched and approached the beast with guns drawn. It was only then that the sea monster was discovered to be the old, discarded inner tube from an automobile tire. Had the bellhop not called the police to investigate, the "sea monster" would have moved on, never to be seen again, and cryptozoology would be discussing the "Miami Beach Sea Serpent."

## The Pablo Beach Problem

Even as Bernard Heuvelmans was finalizing the 1968 English translation *In the Wake of Sea Serpents*, he was receiving new material. One such late arrival was an old page torn from an unknown publication. The engraved image shows two men and two women in vintage swimwear, knee-deep in the ocean and in a panicked retreat from a sea serpent. Heuvelmans offers no text and no identification except for his caption "Excitement at Pablo Beach, 1891."

After Heuvelmans's book went to print, the image appeared regularly in books about sea monsters and sea legends, but always with minimal discussion. No one was able to locate the source or corroborating material. Most recently, Greg Jenkins includes it in his 2010 book, *Chronicles of the Strange and Uncanny in Florida*, including the caption from Heuvelmans and a later caption assigned to it by one of the collections of stock photographs that subsequently included it. He also adds several other captions and then lists his sole source as Heuvelmans, inferring incorrectly that the 1968 book includes the text. Considering Jenkins' book extensively uses material from tabloid newspapers and online social groups, he may not be the most reliable source for citations.

The source of the image remains unknown, but certain details help. The image is now part of the "Bettman Archive," an image library created by Dr. Otto Bettmann in the years immediately before WWII, estimated to be over 11 million pieces at the time of his death. He focused on early photographs and movie stills, but not exclusively. Today, after the collection was sold several times, it is part of the Getty Images portfolio, where the online database notes the original title was "Sea serpent frightens others off Pablo Beach, Florida."

This suggests the caption's wording, "Excitement at Pablo Beach, 1891," is Heuvelmans' own, which also explains the date, which is not correct, based on the swimsuits on the fearful beachgoers. Swimwear was evolving very quickly at the time, and these are not 1891 styles. The bathing suits in the image are from the 1900s time period, not the 1890s.

Heuvelmans may have picked the date hurriedly in the face of publisher deadlines, based on another newspaper clipping in his files, one that did refer to "excitement" on Pablo Beach in 1891. But had he stopped to read the article, this excitement was the tragic story of pranksters who tied a dog to a captured turtle, unaware that the turtle was stronger than the dog, dooming the dog to be dragged with the turtle back into the ocean.

Pablo Beach itself was originally named Ruby until 1886 when it was renamed after the nearby San Pablo River. In 1925, the name Pablo Beach was changed to Jacksonville Beach, an inevitable change expected since 1912 when the San Pablo River ceased to exist after a canal was dug that made it part of the Intracoastal Waterway.

So the image could only use the name Pablo Beach from 1886 to 1925. With the style of the wardrobe factored in, the window can be narrowed down to 1900-1905. Terrified girls in swimwear were just salacious enough not to cause an uproar, suggesting the most obvious source would be *The Illustrated Police News* of London, which gleefully reported and illustrated crimes and scandals. The early tabloid was also perfectly willing to break up the parade of murders and hangings with the occasional damsel in distress story. A similarly themed issue featured a front cover rendering of bathing beauties in the coils of an octopus with the headline proclaiming "Alarming Experience of Fair Bathers Who Are Attacked by an Octopus."

Unfortunately, tabloids were ephemeral and passed among people and literally read until they were in tatters and discarded. Without knowing what Heuvelmans received, we also don't know how accurate his caption is—a transcription error could mean the difference between Pablo Beach or San Pablo Beach. And considering California has swimming on San Pablo Bay (the northern extension of San Francisco Bay), until someone comes across the original source, we do not have the date of publication, the publication's name, or even a confirmation that the location is Florida's Pablo Beach.

## The Soldier Key Skull

Soldier Key is an isolated island in Biscayne Bay, five miles south of Key Biscayne, whose nearest neighbor is the Fowey Rocks lighthouse, three miles east. A fully-staffed, small lodge was built on the key by Henry Flagler as an amenity for wealthy guests of his exclusive Royal Palm Hotel in Miami who wished to pretend they were "roughing it" at a fishing camp. Once the novelty wore off, the lodge was turned into Soldier Key Club, a private fishing club.

Elmer E. Garritson was named the manager, partly an honorific title because he was the first one to arrive and last one to leave each winter. Garritson had parlayed a successful career as the circulation manager at the *New York World* into early retirement to pursue his hobby of fishing and hunting. He was notorious for arriving at the club before it officially opened.

In January of 1921, Garritson was returning to the club when he noticed sharks feeding on something large near the shore. He approached and saw six feet of a massive skull sticking out of the water. With sharks tearing the flesh away, he didn't dawdle, but he estimated that, whatever it was, it was 80-feet long.

On February 12, Garritson used a favor and asked Captain Charles Moller to assist in recovering the carcass. Miller's fishing vessel, the *Corsair,* had a hoist, and between the crew and equipment, they managed to bring the skull to the Royal Palm dock.

It was 15-feet long and weighed three tons. Cartilage hanging off it was as thick as a man's arm. Garritson was still a newspaperman at heart, and the local reporters were his friends and fishing buddies. If Garritson, who held the world record for catching the largest tarpon, didn't know what the skull was, it was obviously an unknown animal. The headlines simply called it a sea monster. The press made sure it hit the Associated Press newswire so Garritson's former colleagues in New York would hear the news. When asked if it could be a whale's skull, Gerritson was adamant it was not: the configuration was completely different. His theory was that it was a huge squid from the depths of Atlantic that had come to shallow water to die.

The "squid skull" remark was enough to bring Louis L. Mowbray into the discussion. Mowbray was the director of the Miami Aquarium and considered an expert in tropical marine life. He was vehement in his denial of the skull's value, probably because the newspapers kept reporting how the skull had scientists puzzled, yet no one had actually asked a marine scientist. He declared that any scientist celebrating the find was not a zoologist. Squid, being a mollusk, have no skull. The skull Gerritson recovered, he said, was the upper part of a whale's.

He closed his interview with the *Miami Herald* with a quote that has become increasingly relevant as time and technology advances: "As director of a scientific institution, I regret the fact that so many erroneous reports are foist upon the public. We are here to give accurate information and are glad to answer any questions as to the identification of marine life to the very best of our ability."

Two weeks later, the *Miami Herald* ran a photo of the "skull of the monster sea serpent." Mowbray's identification may have been casting pearls before swine, but it was a learning experience.

Mowbray would later return to again run the aquarium he started in Bermuda, and when he retired, his son Louis S. Mowbray assumed the job. The younger Mowbray was the director when Sister Jean reported a sea serpent in Hamilton, Bermuda, in 1960 (Chapter 4). The younger Mowbray's approach was less combative and may have helped the nun's encounter fade into the background more quickly than the "sea serpent skull" his father attempted to debunk.

### The Sanibel Monster

In June 1974, Elizabeth Adler was photographing wildlife on the shore when she saw something four-to-six-feet long in the shallow water of the J. N. "Ding" Darling National Wildlife Refuge. She took a photograph and sent it to the local newspaper. The *Island Reporter* immediately recognized the photo as the back of a manatee (*Trichechus manatus latirostris*) but would never risk offending the photographer by just dismissing the matter. Adler was a former Sanibel Island resident who had been active in local politics. She was still active in local charities and was a vocal proponent of protecting Florida wildlife refuges.

Instead, the newspaper ran the photograph with tongue firmly planted in cheek, and asked if their readers could help identify the "Sanibel Monster." Mrs. Adler was apparently offended anyway and shifted her attention to the new Edward Ball Wakulla Springs State Park, 400 miles away (where she would also have encountered manatees).

A second "monster" sight took place in the same place 40 years later. A Boca Raton small business owner and her fiancé posted the footage to YouTube that they shot from a small boat in the shallows of the wildlife refuge. Although they had both encountered manatees before, the behavior was so atypical for the slow-moving behemoths that they believed they saw something different, referring to it as a looking like a 20-foot snake wrapped around some other animal. The animal moved quickly, and broke the surface, briefly thrashing and appearing to be attacking something.

The location confirms the identification as manatees—the refuge has two types of seagrass beds, including manatee grass (*Syringodium filiforme*), commonly found in seagrass beds in warm, shallow waters, and the favorite food of the gentle giants. The behavior is rarely observed but common—the two witnesses saw a manatee mating herd, where multiple male sirenians aggressively jockey for position in an attempt to mate with the female.

The YouTube footage, taken by a camera phone, is brief and at a distance. It is sufficiently vague that the video was circulated on cryptozoology websites with a variety of identifications ranging from a baby whale to an anaconda battling a manta ray, but the reported cryptid attracted no mainstream media interest.

There were several reasons why it was ignored. First was that the current video's description was cleaned up by a representative of a user-generated content media agency that handles licensing the footage, presumably in case a television series was interested. That alone would discourage most newspapers and local news reports, the sort of coverage that would be needed to go viral.

Also, the original video's title was "Did we find Nessie- Clearest and closest Loch ness Monster Video ever caught on camera!!!!!" which sounds more like clickbait than cryptozoology—not that the current title is much better. The original caption also had the fiancé's business URL listed as the first thing you read in the caption, which effectively destroyed the videographers' purported desire for anonymity.

Sanibel Island is a popular tourist destination, known for its beaches filled with seashells. More than half of Sanibel comprises wildlife refuges that provide a safe haven for a wide variety of animals. It does not, however, include sea monsters on the official inventory of species.

## Normandy Nessie

In April 2009, retired engineer Russ Sittloh was sitting in his Madeira Beach home when he saw something in the canal behind the house. When it rose out of the water, Sittloh saw that it was brown on top with a mottled brown and yellow underside. It flipped its tail before submerging. Sittloh described the tail fin as being flat, like a lamprey tail. There are no native lampreys in Florida's Gulf waters he could be comparing it to, so Sittloh was using the lamprey's diphycercal tail fin as a general description as kite or fan-shaped. He did not see a head or a dorsal fin. During a second sighting in September, he estimated a girth of 12-15 inches coming out of the water. It convinced him that it was a giant snake such as a python.

Concerned for the safety of swimmers and kayakers in the canal, he posted a letter warning of the beast to the editor on the Tampa Bay newspaper website on October 7. Responses ranged from skepticism to mockery. The fact that Sittloh nicknamed the creature "Normandy Nessie" after Normandy Road, the street his house was on, didn't help. A Sheriff's Office spokesperson dryly noted that no one else in Madeira Beach had reported seeing a Nessie.

The 78-year-old retired engineer did not take kindly to being dismissed. He purchased a security camera and positioned it at the canal to capture evidence. For eight hours a day, he recorded the canal and then reviewed the video. He finally obtained footage of the beast. When Sittloh posted the video online, it went viral. That made the local newspapers finally take notice.

Sittloh's video showed something brown moving along the surface. Another shows something splashing. He placed five videos on YouTube. In one, Sittloh claimed the creature was in the middle of a school of baitfish, did a double roll, and came back "with a mouthful of fish." A biologist with the Florida Fish and Wildlife Conservation Commission said, based on the speed and size, it looked like a large manatee. Sittloh scoffed at the suggestion, claiming he'd swum with manatees and this was not a manatee.

Interest died down in Madeira Beach (not the internet) and there were no further sightings that year, coinciding with the manatee migrations to warmer waters for the winter. The next year, in late August, 13 miles up the coast in Clearwater's Sand Key Park, a vacationing Ohioan family claimed they spotted a 30-foot-long serpent off the shore. Whether it was a line of dolphins in the distance, an oarfish, or a snake of some sort, only one local TV news covered it, and it was little more than their excuse for revisiting Russ Sittloh's Normandy Nessie story a year after his sightings. They found Sittloh now convinced that Normandy Nessie was one or more anacondas, inferring there had been more sightings, which he thought had been deliberately ignored by authorities to avoid hurting the tourism industry. Vernon Yates of Wildlife Rehab and Rescue was also interviewed, and he pointed out he'd captured several dozen anacondas in the Tampa Bay area over the years with the longest measuring about 15 feet long. More notably, anacondas will avoid saltwater whenever possible.

The two sides would never agree on an identity for Nessie. Russ Sittloh died in November 2011, almost two years to the day his Normandy Nessie went viral. Whether it was a manatee, an anaconda, or something else, the debate will continue to rage on, at least as long as his videos remain online.

## Great Horned Reptiles

Among Native Americans, there is a widespread belief in a horned water-beast, usually a serpent of enormous size. Ethnohistorian George Lankford, in his study "The Great Serpent in Eastern North America," notes that among the various Native American groups in the Eastern Woodlands and the Plains, the cosmos is a series of layers, each of which is identified with particular powers. The bottom layer is primeval water, and the world was created specifically to float upon the waters. The mythological creatures that rule these waters are horned and aquatic. Even in modern times, these residents can come to this world, and can occasionally be seen by humans—but only if they choose to be seen.

In 2007, a group of amateur cave explorers discovered petroglyphs on the walls of a cave in the eastern Florida Panhandle. The cave, on undisclosed private property near Florida Caverns State Park in Mariana, has eight carvings on the wall and is only the second example of prehistoric rock art found in Florida. Based on the style of the carvings and artifacts found in the cave, archaeologists believe the images were made in the Late Woodland Period (500 AD - 1200 AD). The petroglyphs are made of fine lines scratched into the soft limestone wall. All but one carving have geometric designs; the eighth petroglyph is a serpent.

The serpent's head is teardrop-shaped with a central depression representing an eye. It is connected to a long single line representing the body. The line curves sharply three times, giving the snake a shape like a letter M. The carving covers roughly 6 square inches. Although archaeologists were excited by the discovery of the carvings as a whole, the serpent offers additional insight in a broader context. Culturally, the Native Americans of the area are part of the same Southwest Culture that includes, among other sites, the Mud Glyph Cave in East Tennessee, where there are a number of serpent images, including one with horns, known as a Horned Water Serpent.

Horned snake folklore is part of the traditional culture among the Seminóles, particularly the Oklahoma Seminóles, who were driven from Florida and assimilated into the Creek culture, primarily between 1827 and 1842. Anthropologist James W. Howard studied the Oklahoma Seminóles with the assistance of Willie Lena, who was raised by culturally traditional grandparents. Lena was able to introduce Howard to much of his ancestral culture still in practice. According to Lena, the scrapings off the horn of the great horned snake are one of the most potent medicines. The snake is much larger than the ordinary snake, blue on top and yellow beneath, with horns of varying hue. It lives underwater in the deepest parts of ponds and lakes and is fully capable of grasping a man and dragging him underwater to consume.

Although the most common form is the Horned Water Serpent, there are variations, such as a giant fish or underwater panther. Lankford argues that although it is referred to as a single creature, a careful reading of the folklore suggests the Horned Water Serpent is really a race of beings with individual members dwelling in large bodies of water. This Great Serpent appears in lore and iconography across the eastern United States from Machias Bay, Maine, to Florida.

Lankford further cites historian Theresa Smith's *The Island of the Anish-naabeg* in which she summarizes previous studies and observes that descriptions are "invariably of a water serpent, usually horned and always of an immense size." Although the Florida petroglyph is not of the horned variety, there is a Floridian variation that also matches Smith's typology—the "horned alligator."

The best example of the horned alligator is the image of one painted on the side of a small wooden box found at the famed Key Marco excavation site on present-day Marco Island. This painted image of a horned alligator comes from the lid of a wooden box discovered by anthropologist Frank Hamilton Cushing in an 1896 Key Marco excavation, a treasure-trove of wooden artifacts preserved by submergence in the marsh. Because the collapsed box panels were found in proximity to earrings, pearls, and carved discs, Cushing dubbed it the "jewelry box."

The box was found in decayed matting containing a bundle of ceremonial adzes, painted shells, and other objects, either sacred or used to create sacred objects. When Cushing reassembled the box, the image of the horned alligator was on the interior, suggesting to him that its proximity to the contents of the box was either imbuing sacredness to the objects or reinforcing its spirituality.

In both the icon and the petroglyph, we do not have the context of the image, so we can't be certain if they represent an oversized figure of legend or something more tangible. Lankford's typology of horned water serpent could also be considered as falling in the "marine saurian" category of Heuvelmans sea serpent classifications. Described as large crocodile-like animals, often with flippers, Heuvelmans theorized it to be a surviving marine dinosaur of the *Thalattosuchian* group that roamed the seas globally from the Early Jurassic to the Early Cretaceous.

Paleontologist Darren Naish politely refers to this theory as "extraordinarily implausible," noting marine tetrapods, in general, are well-represented fossil record, both their rise and decline. Most marine thalattosuchians did not survive into the late Cretaceous. In support, Naish cites French paleontologist Nathalie Bardet whose work on the evolution and extinction of marine reptiles also documents the extinction of other marine reptiles commonly suggested as sea serpents such as mosasaurs, elasmosaurs, and pliosaurs.

Heuvelmans only lists four examples globally, none of which are in the Southeast United States. The "mystery saurian" category of Coleman and Huyghe mentions only two examples, neither of which is from Florida or the southeast, where alligators and crocodiles are common enough that a 15-foot reptile might not raise eyebrows as they might in other locales.

It is interesting to note that trapper Buster Ferrel encountered a horned snake (Chapter 8) that was certainly large enough to grapple with man-sized prey. Considering how universal the horned serpent is among the Native tribes, the question may not be whether this is a folk memory of a cryptid, but rather how many hunters didn't fare as well as Buster Ferrel.

# CHAPTER 16
## Mermaids, Mermen, and Manatees

Even cryptozoologists question whether the mermaid is a cryptid or a folktale. Fortean author Jerome Clark, for one example, finds it hard to believe any sort of flesh-and-blood merbeing exists. Because mermaids have consistent behavior—they are "animals that cavorted frequently in shallow offshore waters"— he concludes that, like any other coastal marine creature, a corpse of one would eventually wash ashore, meaning "we would have more than sightings to document them."

Folklorist Horace Beck considers the mermaid to be "fractured mythology," a belief so old that it could date back to Neolithic beliefs that have been lost to time, leaving the mermaid as a universal folk figure without the original context. Biologist Richard Carrington further notes, "There is not an age, and hardly a country in the world, whose folklore does not contain some reference to mermaids or to mermaid-like creatures. They have been alleged to appear in a hundred different places, ranging from the mist-covered shores of Norway and Newfoundland to the palm-studded islands of the tropic seas."

Art historian Henry John Drewal traces the arrival of European images of mermaids in West Africa back to figureheads on Dutch and other European ships in the pre-colonial era. Because early European travelers were associated with the sea spirits from the 15th century onwards, a new water spirit arose, known as Mami Wata, who evolved into a singular pan-African deity. When Western Africans were forced into slavery in South America and the Caribbean, Mami Wata came with them, creating a rich history of mermaid sightings, simplified and homogenized by the Europeans under a variety of local names in the language of the colonizing nationality. Variations on Mami Wata include references to the Méné, Mamma, and Femme Poisson.

This aquatic femininity also supports the leading candidate for mermaids—the manatee. The name manatee first appears in English in 1555 in Richard Eden's *Decades of Newe Worlde*, the translation of Pietro Martire d'Anghiera's 1530 history of the Spanish explorers, *De Orbe Nouo*. Manatee is derived from the Carib word *manáti* meaning breast or udder, which may be less salacious then it appears at first blush. Manatee breasts are axillary, not pectoral like a human.

## Columbus' Sirenas

In 1493, Columbus was exploring the Caribbean, following the Hispaniola coast (Haiti) en route to the river now called the Yaque del Norte, when he saw three *serenas* (mermaids) rising up out of the waves. Although they had some human appearance, they were not as attractive as expected. He also noted he had also seen mermaids in African waters off Guinea.

Columbus biographer Samuel Eliot Morison is quick to point out that these are obviously manatee (*Trichechus manatus*) and West African dugong (*Trichechus senegalensis*). Anthony Piccolo, a professor of literature, concurs but doesn't blame Columbus' eyesight. Piccolo explains that Columbus, having studying maps and ship logs, was "mentally primed for mermaids." James Powell, a biologist with the Wildlife Trust, notes "there have been times when they come up out of the water and the light has been such that they did look like the head of a person." Powell further observes that when mermaids are expected, even a glimpse of a manatee's back, with no dorsal fin and tail, would create a self-fulfilling prophecy.

Piccolo also addresses the sniggering remarks about sailors' eyesight by pointing out that although slender figures may be in fashion now, the Age of Exploration was also the age of Rubens, the Flemish painter who popularized a heavier, more voluptuous idealized female body.

## The Hipupiára

In 1554, in the Portuguese Captaincy of São Vicente, Baltesar Ferreira, the son of the colony's leader, was awakened by the screams of his father's Native slave. She had been outside and seen a monster on the beach so ugly it must have been Satan itself. Ferreira proceeded to the beach, and to his shock, there was indeed a monster.

The monster saw Ferreira and started back to the water. Ferreira intercepted it and barred its path. The monster then stood upright like a man on its tail fins. Chivalrously now face to face with his unarmed foe, Ferreira thrust his sword into the beast's belly. The resulting geyser nearly blinded Ferreira as the beast attacked with teeth and claws. Ferreira was able to cut the beast again, this time in the head. Weakening from blood loss, the beast turned back toward the ocean. By now, other slaves had arrived, seized the nearly dead monster, and paraded it around the town.

The monster was a water demon, the *hipupiára*, in the local dialect. It was 15 palms long (11 feet), covered with hair, and had silky bristles on its muzzle. The battle was sufficiently riveting that Gândavo devoted an entire chapter to the incident in his 1576 history. One of the editors of the 2008 edition of that history was ethnozoologist Ricardo Martins Valle of Universidade Estadual do Sudoeste da Bahia. Valle believes the animal killed by Ferreira was a South American sea lion (*Otaria flavescens*) that had traveled north outside its normal range. The size appears to justify the claim, as South American sea lions males can exceed nine feet and are known for stress-triggered aggression (such as blocking the path back to the water). Not mentioned was another likely candidate, a large male southern elephant seal (*Mirounga leonina*). Elephant seals take their name from a large proboscis of the adult male, capable of extraordinarily loud roaring noises. Southern elephant seal bulls typically reach a length of 16 feet. And as discussed in the chapter on the St. Johns River, the male sea lions are known to travel great distances to go ashore and molt all of their old fur and epidermal skin.

Michel Meurger, in his study of lake monsters, agrees that this was a mammal. His issue is that the *hipupiára*, or also called the *ipupiára* or *igpupiára*, appears to be a generic catch-all term for aquatic encounters. In the 1580s, Jesuit priest Fernão Cardim described the male *igpupiára* as having the appearance of a reasonably tall man, but with very deep-set eyes; whilst the female had the appearance of a beautiful woman with long hair. It kills its victims by embracing the person, squeezing them so tightly that it "breaks them without tearing them apart." It eats only the eyes, nose, fingertips, toes, and genitalia, "so they are found on the beach looking ordinary except for the missing parts." Ichthyologist Peter J. P. Whitehead of the British Museum similarly questions the account, noting that much of Father Cardim's "naturalist" observations are just recycled, pre-existing fish tales told to the apparently gullible priest. So, it is no surprise that Cardim's description sounds more like a manta ray than a mermaid. His described method of killing is strikingly similar to Antonio de Ulloa's 1784 description of how a manta ray allegedly killed its victims (see Chapter 14), coupled with normal soft tissue damage inflicted by small marine animals, typical of that found on the corpses of drowned humans.

Meurger also notes other *ipupiára* encounters, most of which he considers more likely to be marine mammals, not merbeings, such as Jean de Léry's 1578 account of a Native who was fishing at sea when an *ipupiára* surfaced and put its fin in the boat. The terrified Native chopped off the appendage and found it had five fingers, like a hand. The Native also said the *ipupiára* had a human face. Meurger suggests that the pectoral flipper of the indigenous Boto dolphin (*Inia geoffrensis*), being of a smaller species and having a fingerlike bone structure in their flipper, could be mistaken for hands. He also notes that the "mutilated mermaid" motif is a well-established folktale in Europe, and desiccated "mermaid hands" were a staple in the collections of scholars who eagerly acquired them from returning explorers. This may be the earliest form of manufactured cryptids, the start of the Feejee Mermaid and Jenny Hanniver market.

## John Smith and the Green-Haired Mermaid

Any history of mermaid folklore is sure to include the encounter of explorer John Smith with a green-haired mermaid in 1614, seven years after the encounter with Pocahontas that Smith would perpetually elaborate upon. Smith was exploring the West Indies when he spotted a beautiful woman with green hair swimming gracefully along the shore. "The upper part of her body resembled that of a woman" with "large eyes, rather too round, a finely-shaped nose (a little too short), well-formed ears, rather too long," but "her green hair imparted to her an original character by no means unattractive." Smith began experiencing love at first sight, at least until the green-haired beauty dove beneath the waves, which, to Smith's horror, revealed she had a fishtail. The event is repeated almost verbatim from book to book even today and usually appears in quotation marks, as if quoting Smith's account.

However, there is no record of meeting a mermaid in Smith's copious journals. In fact, Smith was not in the West Indies in 1614. Professor Vaughn Scribner traced the story back to famous French author Alexandre Dumas. Dumas had written the fantasy story "Les Mariages du Père Olifus" that was serialized in 1849 in French with an English edition, "Nuptials of Father Polypus," a few months later. A far cry from his *The Three Musketeers* or *The Count of Monte Cristo*, the story tells of a Frenchman's quest to find a Dutchman who had allegedly sired four children with a mermaid. The reference to John Smith and the mermaid appears in the first chapter as an example of earlier mermaid sightings.

Scribner is quick to note that Dumas didn't create the story; he merely popularized it. Thanks to the reference by France's most popular writer, by 1867 the story was appearing in popular, presumed to be nonfiction, texts such as Landrin's *Les Monstres Marins*.

Scribner was able to continue tracing the story back to Theodore de Bry in 1628. In one section of de Bry's book, he discusses John Smith's journals, then segues into the narrative of another explorer, Richard Whitbourne. Whitbourne did claim that he had seen a mermaid while sailing off the coast of Newfoundland in 1610. Whitbourne was the source of the John Smith encounter. Stengel's *De Monstris et Monstrosis* in 1647 appears to have been the culprit causing the confusion. Either Stengel missed the transition from Smith to Whitbourne or, as Scribner suspects, Stengel deliberately made the change to use Smith's celebrity to attract readers.

Scribner notes that Dumas then tweaked the story again, also to make it more alluring. The original Whitbourne mermaid neither inspired love at first sight nor even had green hair. Whitbourne's description noted she had "round about upon the head, all blue strakes [streaks], resembling hayre, downe to the Necke, (but certainly it was no haire)."

### Benjamin Franklin's Contribution

The April 22-29, 1736 issue of the *Pennsylvania Gazette* also reported a merfolk sighting. The brief article was taken from a newspaper in Bermuda that reported a "Sea Monster" had been spotted recently. The upper body appeared to be a 12-year-old boy with long black hair and the lower half a fish. First spotted on the shore, a crew took to the water in a boat in hot pursuit. Just as the craft was about to drive a fishgig into the creature, its unexpected human likeness shocked the crew into compassion; they could not kill the creature. This particular sighting is so vague that it would be of limited interest in cryptozoological circles but the newspaper's publisher was Benjamin Franklin, so short or not, the article carries some gravitas.

### The van Batenburg Mermaids

In 1797, British physician Colin Chisholm was investigating a yellow fever epidemic. The US capital, then in Philadelphia, had been ravaged by the scourge that seemed to have coincided with a recent influx of Caribbean immigrants. The immigrants arrived in a ship that had originated at the yellow fever-free West African colony of Bolama (an island off present-day Guinea-Bissau) but infected every Caribbean port of call it made. With the epidemiology of the disease still a mystery, Chisolm was attempting to track the outbreaks and find a common denominator.

Abraham Jacob van Imbyze van Batenburg, the governor of Berbice (now Guyana), told Chisholm, during a social call, of the mermaids found in the Berbice rivers, called the *méné mamma*, or "mother of the water" by the local Natives. The governor described the mermaids as resembling a human, either black or tawny in color. The head was slightly smaller in proportion, usually bare, but occasionally covered with black hair. The shoulders were broad and "the breasts large and well formed." The rest of the mermaid was an immense fish with a forked tail. The mermaids were generally observed in a sitting posture in the water, which hid their piscine half until disturbed, when they dove, exposing their tails and agitating the water for a considerable distance.

Chisolm was dubious, but the doctor knew how to play the diplomacy game. The governor seemed pleased that he would be quoted in the book, so Chisolm carefully reported the "mermaids" as Governor van Batenburg described them, and offered "corroboration" in an additional account in a book by John Stedman. With accessibility to books limited, few readers would realize that Stedman does describe the creature and then identifies it by name as the manatee.

### The Schooner *Addie Schaefler*

The news came out of Jacksonville on April 19, 1890, and spread like wildfire. By September, the story had appeared across the globe—a mermaid had been captured. W. W. Stanton, mate of the schooner *Addie Schaefler*, had been fishing for bass 300 miles off St. Augustine when something got entangled in the line. The creature was about six feet long, pure white, and smooth-skinned. The head and face were human in shape with well-defined shoulders, flippers, and two fluke tail. When it was dragged aboard the ship, it gave a low, moaning cry, sounding like the sobbing of a baby.

Stanton kept the "strange fish" on the *Addie Schaefler*, charging admission to see it as the schooner made its cargo route into the Caribbean. The beast barely survived two days, but even the corpse was a profitable attraction, so much so that Stanton could afford to place the remains in a six-foot glass container of alcohol to preserve it. At the time of the article, Stanton was planning on continuing to exhibit the remains as the schooner continued its route, and then have it shipped to the Smithsonian. Unsurprisingly, the Smithsonian is still waiting.

In this particular case, the identity of the "mermaid" is more intriguing than the siren itself. The article, counting tail flukes as fins, specifically references four fins—the creature has no dorsal fin. The lack of a dorsal fin, the color, and the "well-defined shoulders" can only be one creature—a young beluga whale (*Delphinapterus leucas*). It is also known as the white whale, the only cetacean usually white, or as the sea canary, due to its high-pitched cry. The sudden tapering to the base of the neck of a beluga gives it the appearance of shoulders, a unique feature among cetaceans.

The problem with a beluga whale off the coast of Florida is the location. The beluga is indeed white and without a dorsal fin, but these are adaptations for life as an Arctic and sub-Arctic cetacean, not a creature of the tropics. Vagrant belugas have been spotted as far south as New York, but St. Augustine is another 1000 miles south. This would indeed, as the newspaper article declared, be the strangest fish that has ever been caught.

## Mr. Carruther's Mermaid

The *Brooklyn Daily Eagle* ran an article in early October of 1894, recounting a yarn spun by successful real estate broker and former sailor Frederick Carruthers. Carruthers recalled standing on the deck of an anchored barque off Demerara in British Guiana (present-day Guyana). He was facing the bow, taking a sponge bath. As he drew a bucket to douse himself, he observed a dark-skinned mermaid on a rock about 30 thirty feet away:

> "The face was quite round, a perfect human face of small size, and the hair hung far down the back [shoulder-length]. The color was a little darker than my hand [a deeply tanned Caucasian]. The neck was human, and the shoulders somewhat resembled the shoulders of a woman, but the body rapidly became fish, with big fins and a regular fish tail."

Carruthers called to the cabin boy, but the sound of his voice alarmed the creature and disappeared. By the time the cabin boy arrived, the mermaid was gone.

The story was a light-hearted piece, gently mocking Carruthers for keeping his sailor's propensity for stories. The story, geared toward being a local interest piece, was picked up by other newspapers, who immediately cut the original article in half, removing all the affectionate ribbing that provided context. By the time *The Ottawa Journal* ran the Carruthers article on October 27, 1894, it was a straightforward report of a mermaid. The question is: Did Carruthers see a mermaid on the rocks, or did his soapsuds-impaired vision see a seal sunning itself?

## Captain Lowe's Mermaid

Author, editor, and historian Marie Cappick was a fixture in Key West. Descended from one of the oldest families, she was an expert on the colorful history of the island city. She also worked as the society editor of *The Key West Citizen,* contributed to the *Havana-American News,* and wrote a weekly local history column for the *Florida Keys Sun,* there wasn't much in Key West, past or present, that she wasn't aware of. The exception was one rumored story that both intrigued her and eluded her.

So, after years of searching, her success was triumphantly stated in the headline of her article: "Three-Year Search for Man Who Saw Mermaids Ends on Porch of Key West Dwelling." The man in question was 87-year-old retired sea captain Samuel Lowe. And over 70 years later, he was still not sure what he saw. He just didn't want to call it a mermaid. Needless to say, that's exactly what the headlines read. The article ran in several of the newspapers Cappick contributed to, but when Captain Lowe died in 1934, she published the expanded version of the newspaper article with additional details.

Samuel Lowe was born and raised on Green Turtle Cay, a barrier islands off mainland Great Abaco. Lowe was a sixth-generation descendant of British colonist Matthew Lowe, a founding father of Abaco (and sometimes pirate).

As a youth, Lowe had a skiff, a shallow, flat-bottomed open boat used for coastal fishing, propelled by pushing against the sea bottom with a pole. Lowe recalled that as a teen, he and Edwin Saunders poled over to a place known locally as "Conch Rocks," a small grouping of eight-foot-tall rocks in a dangerously shallow area several miles off Clifton Bluff on the southeast shore of New Providence island. This is the first error in the octogenarian's account. Poling from Green Turtle Cay, on the south side of New Providence, would require traversing over 150 miles, most of which was too deep to pole. Since both Lowe and Saunders had relatives on New Providence, it's probably allowable, after seven decades, to assume Lowe had forgotten they were on that island and decided to go fishing with a borrowed boat.

Lowe and Saunders proceeded along the coast until they reached Clifton Bluff. Because the waters were shallow enough to be a navigational hazard to larger ships, it meant good fishing for a small boat. Lowe recalled the day was bright with not a cloud in the blue sky, and they could see for miles.

Conch Rocks is on the edge of what is called the "tongue of the ocean," a region of deep water in the Bahamas separating the islands of Andros and New Providence. This is the same 6000-foot deep area that Forrest Wood would propose as a likely environment for giant octopuses (Chapter 5) and where the *Alvin* submersible would purportedly see a Plesiosaur (Chapter 4).

As the teens came into view of Conch Rocks, they noticed something on one of the taller rocks near the drop-off into the depths. Well aware this was a dangerous area for a small boat, they positioned the boat within 20 feet of the rocks to get a better look. It was a woman, miles away from dry land, and with no signs of wreckage. Saunders was convinced it was a corpse washed up on the rock.

Lowe wasn't sure who or what she was, but he thought she was sitting waist-deep in the water on the rock and moving. In spite of Saunders' increasing fear, Lowe poled in a little closer. Her hair was green-black and looked like seaweed. When she threw her arms over her head, Lowe was fascinated to see that her arms were flippers like a turtle—she had no hands or fingers. The creature chose to ignore the boys as they shouted and waved, continuing to gaze seaward. Lowe wanted to go closer, but Saunders was nearly prostrate with fear. He remained convinced it was a corpse based on the being's unnatural color.

Lowe had misgivings leaving the potentially stranded woman behind but had the clear impression that she had no interest in being rescued or chatting with a pair of dumbstruck adolescents. Saunders remained so distraught the topic was not discussed on the trip back to shore. That evening, when Lowe told the story over dinner, his mother suggested that the woman was a mermaid and suggested that Sam avoid that area in the future.

Lowe admitted to Cappick that as an adult, even as he sailed bigger and better vessels, he would frequently sail in the Tongue of the Ocean near Conch Rocks just to see if the woman would reappear, but never saw her again. And perhaps in a telling conclusion to Captain Lowe's tale of the enduring call of the siren, he suggested that Cappick track down Saunders to validate the story. But Saunders had been so shaken by the encounter, he and Lowe had drifted apart. Lowe was unaware his childhood friend had died seven years early in a hospital near the house where Lowes was being interviewed.

**An Atlantean Mermaid**

In 1968, a diver named Robert Froster was diving off the coast of Florida, a deliberately vague location, because Froster was searching for the stone structures off Bimini Island that Dr. J. Manson Valentine had claimed was evidence of Mayan settlements in the Bahamas. Later that year, Valentine would announce additional nearby discoveries focused on the "Bimini Road" as proof of the lost city of Atlantis (not Mayans), as predicted by Edgar Cayce decades before.

Both he and Valentine were Florida residents and scuba divers. So even though Valentine had deliberately kept the location of this first stone structure a secret, the scuba diving community was still small and close-knit enough in 1968 that secrets were hard to keep. Froster knew roughly where to look and arranged for a friend to join him.

At the last minute, his dive buddy canceled, but Froster, breaking a cardinal rule, decided to dive alone. As he swam along in search of the "Mayan ruins," he noticed something in his peripheral vision. Something was stirring up the bottom sediment. In the now murky water, he saw something heading toward him with an undulating motion. When it got to within 20 yards, the Australian diver noticed that the creature had arms that appeared to be tipped in talons, and it was reaching out to him. As it drew closer, Froster saw it was a mermaid. Her top half had skin, breasts, and a head of hair, while the other half was piscine. She did not appear friendly, and the increasingly frightened diver would recall her gaze as being nothing short of hateful. Froster chose discretion over science and raced to the surface and boarded his vessel without further incident. Froster would later recall that he felt the mermaid had planned to ambush him, and he would be dead if he hadn't noticed the sediment being stirred up.

When the story appeared in the 1989 book *Great Southern Mysteries*, there was a typo, and the book claimed the incident took place in 1988, not 1968. Author E. Randall Floyd corrected the mistake when he reprinted the story on September 16, 1990, in his syndicated weekly newspaper column, also named "Great Southern Mysteries." But in the age of Google Books and Amazon previews, the damage was done—to this day, accounts of the Froster mermaid encounter proliferated across the internet, all citing it as a 1988 event.

### Water Apes

Bernard Heuvelmans, in a 1986 paper, suggested that the sheer number of merfolk sightings suggested a still-unknown species of *Sirenia* beyond dugongs, manatees, and sea cows. He was, however, willing to consider another remote possibility—an early primate that escaped extinction and adapted to sea-life. This theory was pursued by Mark A. Hall, who theorized that a primitive primate, such as *Oreopithecus bambolii,* had survived extinction in the Miocene epoch by evolving into an aquatic species with webbed hands and feet. Hall further conflates these aquatic apes with mermaids and reports of swamp-dwelling lizard men. Considering that *Oreopithecus* was slightly smaller than chimpanzees, developed in isolation on Italian islands, and its bipedalism is still debated, it is a stretch to consider it a candidate for either a six-to-eight foot tall swamp monster or a manatee-esque coastal dweller.

Coleman and Huyghe also classify merbeings as a category of primates that includes mermaids, lizardmen, and chupacabras, not cryptids typically considered primates. As such, they discuss in their *Field Guide to Bigfoot and Other Mystery Primates* (as opposed to a more intuitive placement in their *Field Guide to Lake Monsters*), a decision they admit in the revised edition foreword was ridiculed.

Coleman and Huyghe break merfolk into two groupings. The marine category is distinguished by a fin appendage, as opposed to the freshwater category with its characteristically angular foot and three-pointed toes. In this classification system, the freshwater varieties are more mobile and prone to coming onto land where they exhibit aggressive behavior. The flaw in classifying merbeings by environment is that it groups diverse sizes and descriptions. Even Coleman notes that merbeings vary in height from dwarf to man-sized and have smooth skin or short fur, or fur that is mistaken for reptilian scales. Regardless if considered a primate cryptid instead of a marine cryptid (much as in Heuvelmans' original classification system), it appears that the merfolk category has such a variety of variations that even more subcategories may be required, along with the hairsplitting needed to place a specimen in a given subcategory.

Merfolks may just be too expansive of a category for a cryptid as widely recognized and Disneyfied as the mermaid. Even Edvard Eriksen's iconic statue *The Little Mermaid* in Copenhagen has over a dozen replicas across the globe. There are hundreds of mermaids in public art display. In Florida, mermaids perform daily, 365 days a year at Weeki Wachee Springs State Park, a vestige of early Florida tourism. You can also rent a kayak or canoe at the park and proceed up the Weeki Wachee River and see manatees. That way, you can claim you've seen a mermaid, regardless of which origin you prefer.

# EPILOGUE

Still more sea monsters are roaming the seas off the Southeast coast, at least according to the internet and various reference books. While fascinating, they are usually not cryptids. Nonetheless, some of them appear to be sea monsters, at least initially. *The Gainesville Sun's* fishing columnist Richard Bowles recalled an event from one of his Gulf trips that, even as a master teller of fish tales, he has hesitated to share for several years.

Bowles and his wife were fishing in the Gulf, 12 miles off Cedar Keys, when his wife noticed "a creature of frightening dimensions" rising to the surface. It was reddish-brown, and the surface reflected the sunlight "as though sequins covered it." Bowles continued that it was "amorphous, with no distinct shape, its margins roughly square, more than 15 feet across." The veteran sports fisherman had no idea what it could be, and as they prepared to start the engine and avoid the creature, it slowly sank again. Then it rose again, this time directly under the boat. Bowles stabbed at the monster with a pole, only to discover it was not a creature. It was a tightly packed school of some unknown fish, which was finally identified by the National Marine Fisheries Service as a squid mass, which the lab had never seen on the surface before. Based on the color and size, it was possibly a mass of Atlantic brief squid (*Lolliguncula brevis*), a four-inch-long squid, reddish-brown, and an inhabitant of warm shallow waters.

The "sequins" had been the thousands of squid eyes reflecting in the sun. Bowles ended the account by wondering how many other odd encounters were never reported for fear of ridicule.

Translation issues have also turned many an ordinary beast into a cryptid. Eberhart's book of cryptids includes the Wihwin of the Mosquito Coast of Honduras and Nicaragua. In his *Mysterious Creatures,* Eberhart simply calls it a sea monster, notes a few traits, and cites Hubert Howe Bancroft's 1875 *Native Races of the Pacific States of North America.* Considering the Mosquito Coast is in Central America on the Atlantic coast, Bancroft is nothing if not flexible. Bancroft's retells the native tales about the Wihwin, "a spirit having the appearance of a horse, with tremendous teeth to devour human prey" that "haunted the hills during the summer, but retired with the winter to the sea." Bancroft added a footnote to the passage because of the use of the term horse, noting that since horses arrived with the Spanish, the reference was either recent or meant another animal (he suggested a deer). The pertinent elements of the Wihwin are that it is four-legged, has a mouth filled with teeth capable of eating human-sized prey, and covers a broad travel range. But that description is not a sea monster; it is an American crocodile (*Crocodylus acutus*). Even today, the crocodile is found far up the Rio Coco, the border between Honduras and Nicaragua at the heart of the Mosquito Coast.

Invasive species have been carried to Florida and the Caribbean, often as hitchhikers on ship cargoes or imported as ill-advised solutions to pest control. However, economic and environmental impacts aside, they are not of interest to a cryptozoologist. The exception is snakes, which present a different issue. It is often difficult to differentiate a report of a "sea monster," an exotic, invasive snake, and a native serpent in a reported sighting. Observations regarding differences in color, size, and aquatic/amphibious nature tend to blur for a witness when an unexpected run-in with something 8-to-15-feet long swimming in the water results in panicked ophidiophobia. And although there have always been reports of oversized snakes, particularly in Florida (Chapter 8), the invasive constrictor snakes blur the line between mundane and monster.

In the 1970s and 80s, exotic pets became a popular and affordable hobby. Exotic snakes were particularly in demand, at least until pet owners discovered snakes like the Burmese python (*Python molurus*) got very large very quickly. And maintaining a 10-foot snake gets expensive almost as quickly. Too often, an exotic snake owner discovered their pet was too big to keep caged, too costly to feed, or too dangerous to keep. Since no one wants to adopt a massive snake, they were released into the wild without regard to the impact on the environment. Traveling zoos were also notorious for misplacing their collection (for instance, the Charlotte County sea serpent in Chapter 10). But as Florida found out the hard way, there is nothing more efficient for dispersing giant snakes into the wild than a category five hurricane. Hurricane Andrew struck the coast south of Miami in August 1992, and among the swathe of devastation, at least 300 pythons escaped from a destroyed private reptile breeding facility in Homestead. By 2013, the Burmese python had proven itself particularly successful at establishing a breeding population in the Everglades and have decimated native wildlife. Estimates range from 10,000 to 100,000 pythons infesting Everglades National Park, Big Cypress National Preserve, and other undeveloped wetlands. Add a few other species that have also been captured in the Everglades, such as reticulated pythons (*Python reticulatus*), Northern African python (*Python sebae*), Southern African pythons (*Python natalensis*), amethystine pythons (*Morelia amethistina)*, and green anacondas (*Eunectes murinus*), and it's a wonder there are any native species left in the Everglades. The state of Florida, which did not regulate the import and keeping of invasive snakes until 2010, did not help.

A snake, regardless of its size or origin, is recognizable. Native species also show up on cryptid lists, misidentified through confirmation bias. All three chapters in this book (9-11) that discuss freshwater cryptids eventually include the sturgeon. There are three species of sturgeon in Floridian waters: Atlantic sturgeon (*Acipsener oxyrinchus oxyrinchus*), Gulf sturgeon (*Acipenser oxyrinchus desotoi*) and shortnose sturgeon (*Acipenser brevirostrum*). All three are anadromous, living in saltwater but traveling into freshwater to spawn. They can grow up to 300 pounds, and anglers rarely catch them because their diet is primary bottom-dwelling, such as larvae, mollusks, and crustaceans. So they are big and unfamiliar "armor-plated" creatures.

Another native species often overlooked to explain a monster sighting is the Alligator Gar (*Atractosteus spatula*). The alligator gar can weigh up to 350 pounds and reach 10 feet in length. It is almost certainly responsible for at least the encounter at Crooked Lake (Chapter 9), which is described variously as having an alligator-like snout and looking prehistoric. It may be responsible for other cryptid accounts, but it is rarely offered as a possible answer. Author Mark Spitzer feels this may not be an oversight since the gar "would give it away" by providing a probable identity that is not cryptid. This deliberate oversight applies to other cryptids, such as the reports from in Lake Norman.

But the one fish that is almost always called a sea serpent is the oarfish (*Regalecus glesne*). The oarfish is so unusual and rare that even recent papers will use the terms "oarfish" and "sea serpent" interchangeably and for a good reason. The fish has a thin, ribbon-like body and can reach lengths exceeding 55 feet. The body is silvery and sports a long red dorsal fin that runs the length of the body, ending with a crest on top of the head; its red pelvic rays rotate like oars of a rowboat when it swims, giving it its name. The rarity of oarfish is why it remains an understudied species, in spite of first being described in scientific terms by Danish naturalist Morten Thrane Brünnich in his 1771 teaching guide for zoology. Ichthyologist Tyson Roberts, considered the foremost authority on oarfish, records only 561 specimens reported over the last three centuries, most of which did not survive to be acquired by museums.

Oarfish have washed ashore across the Caribbean and Florida. A stranding in January 1860 in Hungary Bay, Bermuda, was carefully measured by London barrister John Matthew Jones, a member of the Linnæan Society who had published a book on Bermudian zoology the previous year. He concluded this was an oarfish under its original *Gymnetrus* classification and was obviously a juvenile of the same species seen by the HMS *Dædalus* in 1848, the sea serpent sighting that launched the modern era of sightings. *Harper's Weekly*, for their coverage, reproduced a rendering of the beast as drawn in Bermuda. It is unquestionably an oarfish, but not what the crew of HMS *Dædalus* saw—the description of the ship's cryptid noted it was dark brown, with yellowish-white about the throat, not silver with a bright red dorsal fin.

The Gulf Coast of Florida seems particularly prone to finding this "sea serpent" with three oarfish found in March 1915 alone, first in Indian Rocks, then the next month, 100 miles down the coast, another washed ashore, a 24-foot "sea serpent" near Boca Grande, and then another at nearby Punta Gorda Point. Roberts notes these strandings can occur in clusters, but the reasons remain vague, possibly as a response to environmental conditions.

There have not been any oarfish found in Florida since 1992, but California had a cluster of strandings in 2013; this is a different species, *Regalecus russelii*—just as long, and just as prone to evoke the wonder of sea serpents. The Pacific species has its own folklore that has nothing to do with the lore of sea serpentry. Japanese folklore says that when oarfish come to the surface, a major earthquake is imminent, which seismologist Yoshiaki Orihara reported was "a superstition attributed to the illusory correlation between the two events." While bad news for earthquake prediction, it is probably good news for California, considering how many oarfish that coast has seen.

But the biggest culprit in creating unintentional cryptids is the human brain. Rupert T. Gould referred to this phenomenon in his 1976 book on Loch Ness as "expectant attention," a tendency of people wanting or expecting to see something, only to be misled by something else that their brains interpret as having some resemblance to it. The classic example is the wake of a boat creating long waves that cast shadows resembling an undulating serpent. It is a psychological phenomenon known as pareidolia. While the endless parade of religious iconography seen on toast and in water stains has proven a perennial favorite in the media, pareidolia also has its place in cryptozoology.

In 1912, passengers on a boat bound to Pass-a-Grille on the southern end of St. Pete Island were startled to see a sea serpent in the distance. It had an odd, almost serpentine head, and about 15-to-120 feet behind the head, waves were breaking over the body just below the surface. As the ship neared the beast, it turned out to be a double-crested cormorant (*Phalacrocorax auritus*) sitting on a post with waves splashing on a sand breakwater. For several minutes, however, there was a ship with a passenger list filled with sea serpent witnesses.

Similarly, travel writer Nevin Winter recounts a story from 1900 of a naturalist standing on the beach on the Treasure Coast of Florida. He saw "a gigantic sea serpent disporting itself along the crest of the waves." As he stood in amazement, the serpent turned into a long line of brown pelicans (*Pelecanus occidentalis*) skimming the surface of the water. The story had a happy, non-cryptid ending. The naturalist was Frank Chapman of the American Museum of Natural History, who followed the birds to an island in the Indian River and discovered it was one of the last surviving rookeries on the East Coast. He led the effort that led to the island being designated the Pelican Island National Wildlife Refuge.

Similar pareidolic cases involve manatees, dolphins, even inanimate objects. Thomas Helm, in the same book where he reports on the St Andrew Bay Cat Serpent (Chapter 12), sheepishly tells of a camping trip on Cape San Blas where after a night storm, he and a companion were convinced they were following a sea serpent along a foggy coast. It had a 15-foot neck, a horse-like head, and a hump 20 feet behind it. The two men followed it as it skirted the coastline, occasionally dipping its head down, making puffing and blowing noises. At daybreak, they returned to try and photograph the beast. It was then they discovered that what they had been tracking was a tree that had toppled in the storm and had drifted and bobbed in the tide. The neck and head were part of the root system. The noise, Helm decided, must have been porpoises feeding on fish driven close to shore in the storm.

In their book *Lake Monster Mysteries: Investigating the World's Most Elusive Creatures*, noted skeptics Benjamin Radford and Joe Nickell similarly believe that the famed Mansi photo of the Lake Champlain Monster is also a log, going so far as to recreate the conditions under which the photo was taken. They conclude the area where Mansi took the picture is too shallow for a large creature, and extrapolate that a three-foot length was more likely than the estimated six-foot neck of the object in the Mansi image. Joseph Zarzynski, the most ardent researcher of the Lake Champlain monster, notes that eyewitnesses are not the most reliable resource, politely suggesting "many estimates of length tend to be overstated." Radford and Nickell also make a note of a letter to the editor in *Fortean Times* by Jerry Monk. Monk, who studied Marine Biology at the University College of North Wales, Bangor, was a hydrographic surveyor, the science devoted to measurement and description of features that impact maritime navigation and related industries. Monk, in precise terms, explained how a tree reacts to immersion in water, eventually sinking and then rising again as methane increases its buoyancy. The tree will move around as it finds its balance with its buoyancy center, then it will gradually submerge as the methane leaks out, doing "exactly as described by Mrs. Mansi." Nickell is quick to defend Mansi, whom he believes is completely sincere in her observations. That means the Mansi photo is not a hoax, just pareidolia.

There will still be sea monster sightings. There will also be hoaxes and sincerely believed misidentifications. The question is: What creatures do the ocean depths still hide?

Cryptozoology continues to lead the search.

# ACKNOWLEDGMENTS

Christopher Balzano, *Tripping on Legends* podcast
Joshua Buhs, *From an Oblique Angle* blog
Caroline Chevallier, Le Havre, Bibliothèque municipale
Loren Coleman, International Cryptozoology Museum
Ellen Crain, W. A. Billingsley Memorial Library, Newport, AR
Bobby Derie
Rob Dyke
Fairlynch Museum & Arts Centre
Dr. Joseph F. Gennaro
Mikhail Gershtein
Sarah Glenn, Mystery and Horror, LLC
Joseph Goudsward, National Weather Service
Jerry Hamlin, dinofish.com
Kenneth V. Heard
Gary S. Hatrick
Diane Kachmar, Florida Atlantic University
Stephanie Kania-Beebe
Paul LeBlond
Lexington County Public Library, SC
Terran McGinnis, Marineland Dolphin Adventure
Bruno Mancusi
Scott Mardis
Scott Marlowe
Judy Padgett Moore
Chris Perridas
Vance Pollock
Peter S. Rawlik, South Florida Water Management District
Michel Raynal
Bob Rickard
Robert Robinson, Legend Trippers of America blog
Mike Ross
Ben Speers-Roesch
Michael Swords
Victor Toro

# BIBLIOGRAPHY

Coleman, Loren, and Patrick Huyghe. *The Field Guide to Lake Monsters, Sea Serpents and Other Mystery Denizens of the Deep.* NY: Jeremy P. Tarcher/Putnam, 2003.

Eberhart, George M. *Mysterious Creatures: A Guide to Cryptozoology.* Santa Barbara, CA: ABC-CLIO, 2002.

Heuvelmans, Bernard. *Dans le Sillage des Monstres Marins.* Paris: Plon, 1958. Released in English as *In the Wake of the Sea-Serpents.* London: Hart-Davis, 1968.

Meurger, Michel, and Claude Gagnon. *Lake Monster Traditions: a cross-cultural analysis.* London: Fortean Tomes, 1988.

Naish, Darren. *Hunting Monsters Cryptozoology and the Reality Behind the Myths.* London: Arcturus Publishing, 2017.

Oudemans, A. C. *The Great Sea-Serpent.* Leiden, Netherlands: E.J. Brill, 1892.

Shuker, Karl P. N. *Still in Search of Prehistoric Survivors: The Creatures That Time Forgot?* Greenville, OH: Coachwhip Publications, 2016.

**Chapter 1 - Living Fossils**

**Basking Sharks**

Holder, J. B. "Sea Serpent: The Great Unknown." *The Century Magazine,* June 1892.

Kuban, Glen J. "Sea-monster or Shark? An Analysis of a Supposed Plesiosaur Carcass Netted in 1977." *Reports of the National Center for Science Education* 17, No. 3 (May/June 1997).

Landers, Mary. "Scientists try to ID 'sea monster.'" *Savannah Morning News.*

March 28, 2018.

Roesch, Ben S. "A review of alleged sea serpent carcasses worldwide. (Part two - 1881-1891)." *The Cryptozoology Review* 2, no. 3 (Winter-Spring 1998).

Shuker, Karl P. N. *Extraordinary Animals Revisited*. Bideford, North Devon: CFZ Press, 2007.


**Frilled Sharks**

"Captured by a Sea Serpent." *Pittsburgh Daily Post*. Aug 20, 1896.

[divers encounter a monster]. *Punta Gorda Herald*. April 8, 1920.

"Frisky Sea Serpent." *The Wichita Daily Eagle*. September 18, 1896.

Garman, Samuel. "An Extraordinary Shark." *Bulletin of the Essex Institute* 16 (1884).

"Monster Sea Serpent." *Ocala Evening Star*. August 24, 1896.

"Sea Serpent Seen by Tampa Mariners" *Pensacola Journal*. February 26, 1905.

"Sea Serpent was seen in the Bay." *Miami Metropolis*. July 15, 1908.

Stead, David G. *Sharks and Rays of Australian Seas*. [Sydney, Australia]: Angus and Robertson, [1963].

"Well Known Boatman after a Sea Serpent." *Miami Metropolis*. July 24, 1908.


**Coelacanths**

Anthony, Jean. *Opération Cœlacanthe*. Paris: Arthaud, 1976.

Arnold, Rudy. "He Makes Big Profits on a Small Scale." *Mechanix Illustrated* (May 1956).

Ashton, Ray E., and Pamela S. Ashton. *Handbook of Reptiles and Amphibians of Florida. Part Three: The Amphibians*. Miami: Windward Publishing, 1988.

"Capture 'Living Fossil' Fish." *The Science News-Letter* 63 (January 17, 1953).

De Sylva, Donald P. "Mystery of the silver coelacanth." *Sea Frontiers* 12 (May

1966).

"Des cœlacanthes dans le golfe du Mexique?" *Science et Vie* 911 (August 1993).

"'Fish' With Legs Puzzles Experts." *Seattle Daily Times* (January 18, 1948).

Fricke, Hans. "Living coelacanths: values, eco-ethics and human responsibility." *Marine Ecology Progress Series* 161 (December 1997).

———. "Quastie im Baskenland?" *Tauchen* 10 (October 1989).

Fricke, Hans, and Raphaël Plante. "Silver Coelacanths from Spain Are Not Proofs of a Pre-scientific Discovery." *Environmental Biology of Fishes* 61 (August 2001).

Greenwell, J. Richard. "Animal Enigmas - Prehistoric Fishing." *BBC Wildlife* 12 (March 1994).

Hoernlein, Henry G. "Fortean Loss." *Doubt* 27 (Winter 1949).

Myers, George Sprague. "Issac Ginsburg." *Copeia, the Journal of the American Society of Ichthyologists and Herpetologists* 1976, No. 1 (March 12, 1976).

"Pas de coelacanthes dans le golfe du Mexique!" *Science et Vie* 913 (October 1993).

## Megalodons

Gottfried, Michael D., Leonard J.V. Compagno, and S. Curtis Bowman. "Size and Skeletal Anatomy of the Giant 'Megatooth' Shark *Carcharodon megalodon*." In *Great White Sharks: The Biology of Carcharodon carcharias*, edited by A. Peter Klimley and David G. Ainley. San Diego: Academic Press, 1996.

Grey, Loren. *Zane Grey – A Photographic Odyssey*. Dallas, TX: Taylor Publishing Co., 1985.

Grey, Zane. *Tales of Tahitian Waters*. NY: Harper & Bros, 1931.

Helm, Thomas. *Shark! Unpredictable Killer of the Sea*. NY: Dodd, Mead Co, 1961.

Pimiento, Catalina, et al. "Geographical Distribution Patterns of *Carcharocles megalodon* over time reveal clues about extinction mechanisms." *Journal of Biogeography* 43, no. 8 (August 2016).

Purdy, Robert W. "Paleoecology of Fossil White Sharks." In *Great White Sharks: The Biology of Carcharodon carcharias*, edited by A. Peter Klimley and David G. Ainley. San Diego: Academic Press, 1996.

Roesch, Ben S. "A Critical Evaluation of the Supposed Contemporary Existence of Carcharodon megalodon." *Cryptozoology Review* 3, no, 2 (Autumn 1998).

Shuker, Karl P. N. *In Search of Prehistoric Survivor.* London: Blandford, 1995.

———. "The search for monster sharks." *Fate,* March 1991.

"'White Death' - Startling Shark Story." *The Sun* [Sydney Australia]. January 30, 1918.

### Miami Beach Sea Scorpion

Cutway, Adrienne. "Florida fisherman reels in massive – shrimp?" *The Orlando Sentinel*, September 8, 2014.

Fuller, Curtis. "The Ocean Has Them Too." *Fate,* July 1959.

Hall, Mark A. *Natural Mysteries: Monster Lizards, English Dragons, and Other Puzzling Animals.* Minneapolis: Mark A. Hall Publications and Research, 1991.

"Rough Seas Hide Hairy Old Monster." *The Miami News.* March 13, 1959.

"Sea Monster Off the Beach" *Miami Herald.* March 12, 1959.

Sneigr, Denis. "'Sea Monster' Sighted off Miami Beach." *The Miami News.* March 12, 1959.

### Goblin Sharks

Goodhue, David. "Shrimper lands, releases rare goblin shark." *Florida Keys Keynoter*, May 7, 2014.

Parsons Glenn. R., G. Walter Ingram Jr., and Ralph Havard R "First record of the goblin shark *Mitsukurina owstoni*, Jordan (Family Mitsukurinidae) in the Gulf of Mexico." *Southeastern Naturalist* 1, no. 2 (2002).

## Chapter 2 - Sea Monster Fatalities

### The McClatchie Tragedy

"Chum in Death Presents Window in Memory of Miss M'Clatchie." *Tampa Bay Times*, October 7, 1923.

Coppleson, Victor M. *Shark Attack*. Sydney, Australia: Angus and Robertson, 1958.

"Girl Dies from Attack of Monster Fish While Chum Battles Waves to Save Her." *St. Petersburg Times*, June 18, 1922.

"Mrs. York Injured Climbing Mountain." *Tampa Morning Tribune*, November 9, 1938.

"St. Petersburg Girl Gets Carnegie Hero Medal for Trying to Save Her Chum." *Tampa Morning Tribune*, April 28, 1923.

### The Pensacola Pass Plesiosaur

"4 Boys Still Missing, Search Continues." *The Pensacola Journal*. March 29, 1962.

Armstrong, Jerry. "Series Of Factors Is Cause." *Playground News* (Fort Walton Beach). May 30, 1962.

"Body of One Youth Recovered in Gulf." *The Pensacola Journal*. April 1, 1962.

Brock, Ira. "Battery Cooper Is Scene Of Search." *The Pensacola Journal*. March 27, 1962.

Dinsdale, Tim. *The Leviathans*. London: Routledge & Kegan Paul, 1966.

"Divers Resume Probe of Gulf for Four Today." *The Pensacola Journal*. March 27, 1962.

"Four Missing, Raft is Found." *The Pensacola Journal*. March 25, 1962.

"Gulf Search Resumes." *The Pensacola Journal*. March 26, 1962.

Guthrie, Ken. "4 Feared Drowned in Gulf, 1 Safe." *The Pensacola Journal*. March 26, 1962.

"Hunt Ends For Youth." *The Pensacola News*. March 30, 1962.

McCleary, Edward Brian. "My Escape from a Sea Monster." *Fate* 18, no. 5 (May 1965).

Robinson, Allen. "Grateful Mother Hugs Son." *The Pensacola Journal*. March 26, 1962.

———. "Mother's Prayer Given Answer" *Playground News* (Fort Walton Beach). May 27, 1962.

"Search Again For Four Boys Off Ft. Pickens." *The Pensacola News*. March 29, 1962.

"Searchers Back on Job." *The Pensacola News*. March 27, 1962.

"Search Continues For Missing Boys" *The Pensacola Journal*. March 28, 1962.

"Searchers Hunt for Four Boys off Old Mass." *The Pensacola News*. March 29, 1962.

"Youths Do Not Heed Warning." *Playground News* (Fort Walton Beach). May 27, 1962.

## Chapter 3 - Encounters in the Gulf

Leonard, Ben. "Ink-Credible Find." *Tampa Bay Times*, July 3, 2019.

### The Cuban Leviathan

Imprenta de Ignacio Cumplido, C. Reeditados por la Oficina de Máquinas de la Secretar´ıa de Educación Pública, 1940.

Pinkerton, John. *A General Collection of the Best and Most Interesting Voyages and Travels in All Parts of the World; many of which are now first translated into English*, vol XIII. London: Longman, Hurst, Rees, Orme, and Brown. 1812.

Southey, Robert, and Charles Cuthbert Southey. *The Life and Correspondence of Robert Southey*, vol IV. Longman, Brown, Green, and Longman. 1850.

Thiéry de Menonville, Nicolas-Joseph. *Traité de la culture du nopal, et de l'éducation de la cochenille dans les colonies-françaises de l'Amérique: précédé d'un Voyage a Guaxaca*. Paris: La veuve Herbault, 1787.

## SS *Neptune*

Gautron, Georges. *La Vérité* sur le grand serpent de *mer, par le Cte Georges Gautron*. Rouen: J. Girieud, 1905.

"La Havane. Cétacé inconnu." *Revue Des Deux Mondes.* 2nd ser., vol. 1 (February-March 1830).

"The Sea Serpent Out-Done." *Charleston Courier*, February 18, 1830.


## The barque *Ville de Rochefort*

Hain, James H. W., et al. "Feeding behavior of the humpback whale, Megaptera novaeangliae, in the western North Atlantic." *Fishery Bulletin* 80, no. 2 (1982).

"Le Serpent de Mer." *Journal du Havre,* August 26, 1840. (Le Havre Public Library, PJ 2 - n° 4245 - 1840-08-26).

"The Sea Serpent." *The Standard* [London], September 15, 1840.

"The Sea Serpent." *The Zoologist* 5 (1847).

Stephens, John Lloyd. *Incidents of Travel in Central America, Chiapas, and Yucatan*. London: J. Murray, 1842.


## The barque *St. Olaf*

Magnus, Olaus. *Historiae de gentibus septentrionalibus*. Antverpiae: apud Ioannem Bellerum, 1562.

"Sea Serpent." *The Galveston Daily News*, May 17, 1872.

"Sea Serpent, Lately Seen Near Galveston." *The Graphic - an illustrated weekly newspaper* [London], August 17, 1872.

Ullmann, A. C. "En norsk skibskapteins beretning om sjøormen." *Naturen*, 1893.


## SS *Pecos*

"40-Foot Sea Serpent Collides With Ship and Drops Into Gulf." *The Brooklyn Daily Eagle*, February 21, 1934.

"Capt. Baker Thinks It was Whale Shark and Not Sea Serpent His Vessel Struck." *Galveston Daily News*, March 1, 1934.

"Strange Animal is Struck by Ship." *New York Times*, February 22, 1934.

## SS *Steel Inventor*

Fitzpatrick, William. "Bo's'n Relates Strange Stories Of 'Monsters' Met in Travels." *New Orleans Item*, March 17, 1934.

LeBlond, Paul H., and Edward Lloyd Bousfield. *Cadborosaurus: Survivor from the Deep*. Victoria, British Columbia: Horsdal & Schubart, 1995.

"Real 'Sea Serpent' In Gulf." *New Orleans Item*, March 6, 1934.

## Chapter 4 - Caribbean Cryptids

### HMS *Orontes*

Pycraft, W. P. "The 'Sea-Serpent' Appears to the Scientist." *Illustrated London News*, June 30, 1906.

Yonge, Reginald F. "The Sea Serpent in 1873." *Illustrated London News*, July 14, 1906.

### The Adventures of the RMS *Mauretania*

"2 Serpents, Not One, Seen By Mauretania." *New York Times*, March 24, 1934.

"Serpent Haunts Sailor in Sleep." *The New York Sun*, February 23, 1934.

Williams, T. Walter. "Mauretania Sights Sea Serpent; Entry in the Ship's Log Proves It." *New York Times*, February 11, 1934.

Williams, T. Walter. "Monster Shark Died Weeping, 'Herodotus' of Cruise Ship Says." *New York Times*, April 14, 1935.

Williams, T. Walter. "New Sea Serpent Reported By Ship." *New York Times*, March 10, 1934.

Williams, T. Walter. "Ship's Officer Sticks to Sea Serpent Yarn, But Says

Artist Failed Reproducing Sketch." *New York Times,* February 24, 1934.

## The Nun's Serpent

Kennedy, Jean. *Bermuda Hodge Podge.* Boston: Bermuda Book Store, 1975.

"The Week in Brief." *The Bermuda Recorder,* October 1, 1960.

## The Haitian Water-horse

"Sea Monster Shelled by Coast Guard," *Haiti Sun,* February 26, 1961.

## The Alvin Plesiosaur

Berlitz, Charles. *Doomsday, 1999 A.D.* Garden City, NY: Doubleday, 1981.

Berlitz, Charles. *Without a Trace.* Garden City, NY: Doubleday, 1977.

Edwards, R. L, and K. O. Emery. "The View from a Storied Sub — The 'Alvin' Off Norfolk, Va." *Commercial Fisheries Review* 30, no 8-9 (August-September 1968).

Garcia Párraga, Daniel, Michael Moore, and Andreas Fahlman. "Pulmonary ventilation-perfusion mismatch: a novel hypothesis for how diving vertebrates may avoid the bends." *Proceedings of the Royal Society B: Biological Sciences* 285, no. 1877 (April 25, 2018).

Kusche, Larry. [review of *Without a Trace*]. *The Zetetic* 2, no,1 (Fall/Winter 1977).

Kusche, Lawrence David. *The Bermuda Triangle Mystery, Solved.* London: New English Library, 1975.

Motani, Ryosuke, "Warm-Blooded 'Sea Dragons'?" *Science* n.s. 328. No. 5984 (June 11, 2010).

## Bermuda Sphere

Agassiz, Elizabeth, and Alexander Agassiz. *Seaside Studies in Natural History.* Boston: Ticknor and Fields, 1865.

Lee Jane J. "What is the car-size ball of jelly that mystified divers?" *National Geographic*. August 6, 2015. https://www.nationalgeographic.com/news/2015/08/150806-mysterious-squid-eggs-ocean-animals-science/.

Mangiacopra, Gary. "A Monstrous Jellyfish?" *Of Sea and Shore* 7, no. 3 (Fall 1976).

Staaf, Danna J. et al. "Natural Egg Mass Deposition by the Humboldt Squid (Dosidicus Gigas) in the Gulf of California and Characteristics of Hatchlings and Paralarvae." *Journal of the Marine Biological Association of the United Kingdom* 88, no. 4 (August 2008).

Winer, Richard. *The Devil's Triangle*. NY: Bantam Books, 1974.

## The Monster of San Miguel Lagoon

"Cubans Give Credence to Mythical Sea Monster." *Orlando Evening Star*, November 10, 1971.

McReynolds, Martin. "Cubans Debunk Monster." *Tallahassee Democrat*, August 23, 1971.

Saavedra, Mario Masvidal. "El monstruo de la laguna de San Miguel." *OnCuba News*. May 24, 2012. https://oncubanews.com/informe/el-monstruo-de-la-laguna-de-san-miguel/.

## The St. Lucia Thing

Marsden, Joan Rattenbury. "Polychaetous annelids from the shallow waters around Barbados and other islands of the West Indies, with notes on larval forms." *Canadian Journal of Zoology* 38. (1960).

Roesch, Ben S. "'The Thing': A Cryptic Polychaete of St. Lucia," *Cryptozoology Review* 1, no. 1 (Summer 1996).

Salazar-Vallejo, Sergio I., Luis F. Carrera-Parra1, and J. Angel de León-González. "Giant Eunicid Polychaetes (Annelida) in shallow tropical and temperate seas." *Revista de Biología Tropical* 59, no. 4 (December 2011).

Stearns, Walt. "Close Encounter with St. Lucia's Thing." *Underwater Journal*, August 25, 2007. https://underwaterjournal.com/2007/08/25/.

**The Bermuda Blobs**

Ellis, Richard. *Monsters of the Sea*. NY: Alfred A. Knopf, 1994.

"'Blob' to stay where it is for now." *Royal Gazette* [Hamilton, Bermuda], January 14, 1997.

Kim, Kiho, and C. Drew Harvell. "The Rise and Fall of a Six-Year Coral-Fungal Epizootic." *The American Naturalist* 164, no. S5 (November 2004).

"Mystery blob believed to be part of whale." *Royal Gazette* (Hamilton, Bermuda), June 14, 1997.

"Mystery blob contains killer fungus." *Royal Gazette* (Hamilton, Bermuda), December 24, 1998.

"Mystery over 'blob.'" *Royal Gazette* (Hamilton, Bermuda), January 13, 1997.


**Chapter 5 - Krakens of the Caribbean**


De Montfort, Pierre Dénys. *Histoire Naturelle, générale et particuliere des mollusques, animaux sans vertèbres et a sang blanc.* Paris: F. Dufart, 1802.

Palsgrave, John. *L'esclarcissement De La Langue Francoyse.* [London]: Iohan Haukyns, 1530.

Pontoppidan, Erik. *The Natural History of Norway.* trans. Andreas Berthelson. London: A. Linde, 1755.

Stéenstrup, Japetus. *The Cephalopod Papers of Japetus Steenstrupt.* Trans. Agnete Volsøe, Jørgen Knudsen, and William Rees. Copenhagen: Danish Science Press, 1962.


**The Giant Scuttle**

"Giant Octopus Blamed for Deep Sea Fishing Disruptions." *International Society of Cryptozoology Newsletter* 4, no. 3 (Autumn 1985).

Greenwell, J. Richard. "Interview [with Paul H. LeBlond and Forrest G. Wood]." *International Society of Cryptozoology Newsletter* 2, no. 1 (Spring 1983).

Lindberg, William J., and Elizabeth Lewis Wenner. *Geryonid Crabs and Associated Continental Slope Fauna: A Research Workshop Report.* [Gainesville, FL]: Florida Sea Grant College Program, Technical Paper 58. 1990.

Luckhurst, Brian. "Discovery of Deep-water Crabs (Geryon spp.) at Bermuda – a New Potential Fishery Resource." *Proceedings of the 37th Annual Gulf and Caribbean Fisheries Institute,* 1984.

Manning, Pickford, Grace E. "Le Poulpe Américain: A Study of the Littoral Octopoda of the Western Atlantic." *Transactions of the Connecticut Academy of Arts and Sciences* 36 (July 1945).

Marsden, Joshua. "A Descriptive Epistle from Bermuda." *Leisure Hours, or, Poems, Moral, Religious, & Descriptive.* New York: Author, 1812.

"Octopus Remains Washed Ashore." *The Freeport News* (Grand Bahamas). January 20, 2011.

Raymond B., and L. B. Holthuis. "Notes On Geryon from Bermuda, With The Description of *Geryon Inghami,* New Species (Crustacea: Decapoda: Geryonidae)." *Proceedings of the Biological Society of Washington* 99, no. 2 (June 1986).

Voss, Gilbert L. "Bermudan Cephalopods." *Fieldiana: Zoology* 39, no. 40 (November 17, 1960).

———. "The Biogeography of the Deep-Sea Octopoda." *Malacologia* 29, no. 1 (1988).

———. "A Checklist of the Cephalopods of Florida." *Quarterly Journal of the Florida Academy of Sciences* 19, no. 4 (December 1956).

———. "Hunting Sea Monsters." *Sea Frontiers* 5, no. 3 (August 1959).

Voss, Gilbert L., and Nancy A. Voss. "An ecological survey of Soldier Key, Biscayne Bay, Florida." *Bulletin of Marine Science* 5, no. 3 (1955).

Wood, F. G., and Joseph F. Gennaro, Jr. "An Octopus trilogy." *Natural History* 80 (March 1971).

"Zoology" [George letter to Wilder]. *The American Naturalist* 6, No. 12 (December 1872).

## Lusca: Beast of the Blue Hole

Agassiz, Alexander. "A Reconnaissance of the Bahamas and of the Elevated Reefs of Cuba." *Bulletin of the Museum of Comparative Zoology* 26. (1894).

Brock, William R. *Scotus Americanus: A Survey of the Sources for Links between Scotland and America in the Eighteenth Century*. Edinburgh: Edinburgh University Press, 1982.

Craton, Michael, and Gail Saunders. *Islanders in the Stream: A History of the Bahamian People*. Athens: The University of Georgia Press, 1999. Palmer, Rob.

*Deep into Blue Holes: The Story of the Andros Project*. London: Unwin Hyman, 1989.

Marsden, Joshua. "A Descriptive Epistle from Bermuda." *Leisure Hours, or, Poems, Moral, Religious, & Descriptive*. New York: Author, 1812.

Palmer, Robert. "In the Lair of the Lusca," *Natural History* 96, no. 1 (January 1987).

Palmer, Robert J., et al. "Appendix I: Inventory of blue hole sites explored or visited on South Andros Island, Bahamas." *Cave and Karst Science - The Transactions of the British Cave Research Association* 25, No. 2 (August 1998).

Révoil, Bénédict-Henry. *Pêches dans l'Amerique du Nord*. Paris: L. Hachette, 1863.


## The Bimini Beast

Davis, Miller. "Mystery Sea Beast Seen Lurking Off Bimini." *Miami Herald*, November 3, 1963.

Fuller, Curtis. "I See From the Papers." *Fate*, April 1964.

Mangiacopra, Gary S. "The Great Ones - A Fragmented History of The Giant And The Colossal Octopus." *Of Sea and Shore* 7, no. 2 (Summer 1976).

## Chapter 6 - The Giant Octopus of St Augustine

"Big Octopus on the Beach." *Florida Times-Union.* December 1, 1896.

Cousteau, Jacques-Yves, and Philippe Diole. *Octopus and Squid, the Soft Intelligence.* Garden City, N.Y: Doubleday, 1973.

Gennaro, Joseph. "Octopus Giganteus: Largest Creature in the World?" *Argosy* 376 (March 1973).

Gennaro, Joseph F., and E. Deborah Elek. "Scanning Electron Microscope (SEM) Disclosure of the Effects of Snake Venom on Living Tissue Integrity and Cell Surfaces." *Scanning* 32 (May-June 2010).

Lucas, Frederic A. "The Florida Monster." *Science* n.s. 5, no.116 (March 19, 1897).

———. "Some Mistakes of Scientists." *Natural History* 28 (March-April 1928).

Mackal, Roy P. "Biochemical Analyses of Preserved *Octopus giganteus* Tissue." *Cryptozoology* 5 (1986).

Mangiacopra, Gary S. "*Octopus giganteus* Verrill: A New Species of Cephalopod." *Of Sea and Shore* 6 (Spring 1976). Reprinted as "*Octopus giganteus* Verrill: Giant Octopus or Whale" in *Fate* 29 (July 1976).

Mangiacopra, Gary S., et al. "An Octopus in the Hand." INFO Journal No. 8 (Winter-Spring 1972).

Moffitt, Donald. "A 200-Foot Octopus Washed Up in Florida, Two Scientists Claim." *The Wall Street Journal* (April 8, 1971).

"Monsters of the Deep." *Arthur C. Clarke's Mysterious World.* ITV, season 1, episode 2, September 9, 1980.

"News of Universities, Museums, and Societies." *Natural Science* 10 (May 1897).

Pierce, Sidney K., et al. "On the Giant Octopus (*Octopus giganteus*) and the Bermuda Blob: Homage to A. E. Verrill." *Biological Bulletin* 188, no. 2 (April 1995).

Pierce, Sidney K., et al. "Microscopic, Biochemical, and Molecular Characteristics of the Chilean Blob and a Comparison with the Remains of Other Sea Monsters: Nothing but Whales." *Biological Bulletin* 206, no. 3 (June 2004).

Raynal, Michel. "The Case for the Giant Octopus." *Fortean Studies* 1 (1994).

———. "Le Poulpe Colossal des Caraïbes." *Clin d'œil* 16 (1987).

———. "Properties of Collagen and the Nature of the Florida Monster." *Cryptozoology* 6 (1987).

Raynal, Michel, and Michel Dethier. "Le 'Monstre de Floride' de 1896: Cétacé ou Poulpe Colossal ?" *Bulletin de la Société Neuchâteloise des Sciences Naturelles* 114 (1991).

Spalding, Frank. "The Facts about Florida." *Miami Daily News Sunday Magazine* (n.d., ca. 1946).

Verrill, A. E. "Additional information concerning the giant Cephalopod of Florida." *American Journal of Science* 4th ser, 3 (February 1897).

———. "A Colossal Cephalopod on the Florida Coast." *American Journal of Science* 4th ser, 3 (January 1897).

———. "The Colossal Cephalopods of the North Atlantic." *American Naturalist* 9 (January 1875).

———. "The Colossal Cephalopods of the North Atlantic II." *American Naturalist* 9 (February 1875).

———. "The Florida Sea-Monster." *American Naturalist* 31 (April 1897).

———. "The Florida Sea-Monster." *Science* n.s. 5 (March 19, 1897).

———. "A Gigantic Cephalopod on the Florida Coast." *The American Journal of Science* 3, no. 13 (January 1897).

———. "The Giant Cuttle-Fishes of Newfoundland and the Common Squids of the New England Coast." *American Naturalist* 8 (March 1874).

———. "The Supposed Great Octopus of Florida: certainly not a cephalopod." *The Annals and Magazine of Natural History,* ser. 6, vol. 19 (June1897).

Verrill, A. Hyatt. *The Ocean and Its Mysteries.* New York: Duffield & Co, 1916.

———. *The Strange Story of Our Earth.* Boston: L.C. Page, 1952.

Voss, Gilbert L. "Hunting Sea Monsters." *Sea Frontiers* 5, no. 3 (August 1959).

Wood, F. G., and Joseph F. Gennaro, Jr. "An Octopus trilogy." *Natural History* 80 (March 1971).

## Chapter 7 - Atlantic Encounters

### The schooner Eagle

[whale carcass on Jekyll Island] *Georgian* (Savannah), July 22, 1930.

[whales in St. Simons Bay] *Georgian* (Savannah), April 22, 1930.

"Sea Serpent in Georgia!" *Charleston Courier,* March 29, 1830.

### The schooner *Lucy & Nancy*

Egede, Poul Hansen. *Continuation af den Grønlandske Mission. Forfattet i form af en Journal fra Anno 1734 til 1740.* Copenhagen: Johan Christoph Groth, 1741.

———. *Efterretninger om Grønland uddragne af en Journal fra 1721 til 1788.* Copenhagen: Hans Christopher Schrøder, 1789.

M'Quhae, Peter, Edgar Drummond, et al. "The Great Sea-Serpent." *The Zoologist* 6 (1848).

"The Sea Serpent Southward." *The Florida Republican* (Jacksonville). March 8, 1849.

### The steamer *William Seabrook*

Murray, Amelia M. *Letters from the United States, Cuba and Canada.* London: John W. Parker and Son, 1856.

"Latest from the Sea Serpent." *Morning News* (Savannah), March 21, 1850.

"The Sea Serpent. Beaufort, March 15, 1850" *Charleston Courier,* March 18, 1850.

"Visit of the Sea Serpent." *Morning News* (Savannah), March 12, 1850.

### The steam packet *William Gaston*

"The Sea Serpent Turned up Again." *The Savannah Republican*. January 17, 1853.

### The *Saladin* and the Balloon Fish

"A Balloon Fish." *The New York Herald*, March 19, 1870.

### Mr. Tuttle's Octopus

"Battle with an Octopus." *New Haven Register*. February 22, 1897.

Mitchell-Hedges, F. A. *Battles with Monsters of the Sea.* New York: Appleton-Century, 1937.

Poli, François. *Sharks Are Caught at Night.* trans. Naomi Walford. Chicago: Henry Regnery Co., 1959. English edition of *Les Requins se Pêchent la Nuit*, 1957.

Roper, Clyde F. E. et al. "A Compilation of Recent Records of the Giant Squid, Architeuthis dux (Steenstrup, 1857) (Cephalopoda) from the Western North Atlantic Ocean, Newfoundland to the Gulf of Mexico." *American Malacological Bulletin* 33 no. 1 (March 2015).

### The Key West Serpent That Wasn't

Meade-Waldo, E. G. B., and Michael J. Nicoll. "Description of an Unknown Animal seen at Sea off the Coast of Brazil." *Proceedings of the Zoological Society of London* 2 (1906).

Sweeney, James B. *Sea Monsters: A Collection of Eyewitness Accounts.* New York: David McKay, 1977.

### SS *Craigsmere*

Coleman, Loren, and Patrick Huyghe. *The Field Guide to Lake Monsters, Sea Serpents and Other Mystery Denizens of the Deep.* NY: Jeremy P. Tarcher/Putnam, 2003.

## SS *Santa Clara*

Ley, Willy. *The Lungfish, the Dodo & the Unicorn: An Excursion into Romantic Zoology*. New York: Viking Press, 1948. The material was included in the compilation *Willy Ley's Exotic Zoology*, NY: Bonanza Books, 1987.

Sanderson, Ivan T. "Mariners Beware." *The Lookout* 55, no. 3 (April 1964).

"Sea Monster Reported Struck by a Grace Liner." *New York Times.* December 31, 1947.

"Tugboat Tows Second Whale into Norfolk." *Richmond Times-Dispatch* (VA). January 3, 1948.


## New Smyrna Beach

Browning, Robert S., [letter to Forrest G. Wood], February 10, 1959. Marineland Dolphin Adventure archives.


## The Giant Juno Worm

"In the Media." *No Bones – the newsletter of the Department of Invertebrate Zoology* (Smithsonian Institution) 21, no. 2 (Spring 2007).

"Scientists Baffled As Florida Diver Captures Sea Serpent on Video; 'Undescribed.'" *Underwatertimes.com*, May 2, 2007. https://www.underwatertimes.com/news.php?article_id=49652103108.

"Sea Serpent off our shore." *NewsChannel 5 at 6*. WPTV, West Palm Beach. May 3, 2007.

Stover, Sarah. "Man Discovers 'Creature.'" *Hometown News* (Palm Beach Gardens), May 18, 2007.


## Chapter 8 - Swamp Serpents of the Treasure Coast


## Juno Ridge

Bell, Emily Lagow. *My Pioneer Days in Florida, 1876-1898*. Miami: McMurray Printing Co, 1928.

## Titusville

"Ah There! – Old Veritas of the Sea Comes to the Surface Again as Natural as Life." *New York Sun,* April 7, 1895.

"Another Fake Story – River Monster" *Florida Star* [Titusville, FL], April 26, 1895.

"Mammoth Sea Serpent." *St Louis Globe-Democrat,* March 25, 1895.

## Okeechobee

"Strange Monster is Killed in Florida." *New York Times,* December 01, 1901.

## The Zimmerman Case

"The Lady or the Sea Serpent – Who Saw Mr. Zimmerman First – and Last?" *The Item-Tribune* (New Orleans), August 19, 1934. (syndicated).

Mutual Life Insurance Company of New York v. Zimmerman, 75 F.2d 758 (5 Cir. 1935).

## Chapter 9 - Lake Monsters

"Goodthing Lake 'Monster' Really is only Alligator." *Pensacola News Journal,* June 9, 1966.

## The Orange Lake Monster

"An Orange Lake Monster." *The Ocala Evening Star,* September 12, 1903.

## Lake Clinch

Beadle, June Freeman. "Scotland has its 'Nessie' and Polk has its 'Clinchy.'" *Winter Haven News Chief,* September 25, 1989.

"C. M. Mallett Death Shock to Friends Throughout State." *Tampa Morning Tribune,* September 11, 1926.

Carr, Robert S., and Willard Steele. *Seminole Heritage Archaeological and Historical Survey: Seminole Sites of Florida.* (AHC Technical Report #74). Miami: Archaeological and Historical Conservancy, 1993.

Hetherington, M. F. *History of Polk County, Florida; Narrative and Biographical.* Saint Augustine, FL: Record Co. Printers, 1928.

Marlowe, Scott C. *The Cryptid Creatures of Florida.* Winter Haven, FL: Pangea Press, 2014.

Ohlinger, Sophronia Carson. "Early Days of Frostproof - Chapter XIV." *Frostproof Highland News*, May 26, 1939.

Stoner Allan. "Effects of environmental variables on fish feeding ecology." *Journal of Fish Biology* 65, no. 6 (December 2004).

Worthington, J. E. "Prehistoric Monster Alleged to Have Been Seen in Crooked Lake." *Tampa Tribune*, October 10, 1922.


**Lake Tarpon**

Henry, Kaylois. "Lost Manatee finds way to Lake Tarpon." *Tampa Bay Times*, October 24, 1992.

McKinney, Matt. "Croc found in Lake Tarpon." *Tampa Bay Times*, July 12, 2013.

Rondeaux. Candace. "Manatee slips in for a tour of Lake Tarpon." *Tampa Bay Times*, August 16, 2003.


**The Lake June Monster**

Belcher, Walt. "Old Moe - Shore Residents Say Lake Monster Lurks" *The Tampa Tribune*, November 15, 1977.

"Catfish Legend Lures July 4th Holiday Anglers." *The Tampa Tribune*, July 2, 1966.

Wing, Ash. "If Lake June Has Monster Catfish, It's Still There." *The Tampa Tribune*, August 11, 1966.


**Lake Zephyr**

Hatrick, Gary S. "Celebrating Celtic heritage in Zephyrbrae." *Zephyrhills News*, January 30, 2003.

Hasselman, Dave. "The legendary tale of Zeffie." *Zephyrhills News,* February 5, 2004.

## Lake Norman

Garitta, Tony. "Lineberger's cat at Lake Norman is a state record." *The Dispatch* [Lexington, NC], July 2, 2004.

Mintzer, Lou. "Looking for big fish? Lake Norman's got 'em." *The Charlotte Observer*, April 20, 2010.

Myers, Matthew. *Lake Norman Monster: A Decade of Sightings*. Mooresville, NC: May 4th Productions, Inc., 2012.

## The Lake Worth Muck Monster(s)

Abramson, Andrew. "W. Palm revives Muck Monster hype." *Palm Beach Post*, October 8, 2010.

Howard, Willie. "Music, 'muck monster' in Lake Worth for Earth Day." *Palm Beach Post*, April 2, 2012.

Miller, Kimberly. "Seal? Manatee? or ..." *Palm Beach Post*, August 16, 2009.

"MonsterQuest: Florida's Cold Weather Reveals Manatee With Strange Tail." *Underwatertimes.com*, January 11, 2010. https://www.underwatertimes. com/news.php?article_id=05329810764.

"Sea Monsters." *MonsterQuest*. History Channel, season 3, episode 10, April 22, 2009.

## Chapter 10 - River Monsters

## The Savannah River Monster

Kirby, Bill. "'Sea Serpent' captured, never identified." *The Augusta Chronicle (Georgia)*, August 29, 2010.

———. "'Sea Serpent' puzzled Augustans in 1820" *The Augusta Chronicle (Georgia)*, August 20, 2014.

O'Neill, J. P. *The Great New England Sea Serpent*. Camden, ME: Down East Books, 1999.

"Sea Serpent." *The Augusta Chronicle & Georgia Gazette,* September 5, 1820.

"Sea Serpent Caught!!!" *The Augusta Chronicle & Georgia Gazette,* September 9, 1820.

**The Needhelp Serpent**

Brown, Loren G. *Totch: A Life in the Everglades*. Gainesville: University Press of Florida, 1993.

"Chokoloskee and Needhelp." *Fort Myers Press*, October 8, 1908.

**Charlotte County Sea Serpent**

"$5,000 Prize Offered to Captor of Seas Serpent (If Real, 30 Feet Long)." *The Tampa Tribune*, November 27, 1962.

"Charlotte County Sea Serpent May Have Six-Foot Babies." *Fort Myers News-Press*, December 29, 1962.

Cortes, Josephine. "Sea Serpent, Treasure: Resident Sees Both." *Sarasota Herald-Tribune*, January 9, 1963.

Farris, Fred. "Charlotte's Sea Serpent May Not be Faring to Well in Cold" *Fort Myers News-Press*, December 14, 1962.

———. "Sea Hunt Fails to Find Snake; May Be a Pair" *Fort Myers News-Press*, December 3, 1962.

———. "Youth Disturbed by Sea Serpent in Myakka River" *Fort Myers News-Press*, November 27, 1962.

Kelly, Patrick. Charlotte, Sarasota Countians Race to Capture 'Sea Serpent.'" *The Tampa Tribune*, November 30, 1962.

"Tice Couple Sights Serpent Swimming in Caloosahatchee." *Fort Myers News-Press*, May 3, 1963.

"What is Doing in Fort Myers." *The Tampa Tribune*, March 11, 1903.

**Altamaha Monster**

Anderson, Bethany. "The legendary Altamaha monster wasn't found off the Georgia coast; here's why it's a hoax." WTLV-TV, March 29, 2018. https://www.firstcoastnews.com/article/news/the-legendary-altamaha-monster-wasnt-found-off-the-georgia-coast-heres-why-its-a-hoax/77-533319231.

Davis, Ann Richardson. *The Tale of the Altamaha "Monster."* Waverly, IA: G & R Publishing. 1996.

Davis, Ann Richardson, and Rhett Davis. *"Sightings of the 'Altamaha-Ha' or river creature of Darien, GA."* Retrieved from http://rhettdavisboy.angelfire.com/sighting.htm (archived site last updated 1999).

Davis, Jingle. "'Monster' Reported in Altamaha River." *The Atlanta Constitution.* Feb 16, 1981.

Floyd, Randall. "Prehistoric creature may live in river." *Strange Encounters* (syndicated column), *Augusta Chronicle.* March 15, 1998.

Magin, Ulrich. "Necessary monsters: claimed 'crypto-creatures' regarded as genii locii." *Time and Mind* 9, no. 3 (September 2016).

Morehead, Bill. "Why I Eat Where I Eat: Higdon's." *The Atlanta Constitution.* June 7, 1981.

Russell, Kathleen. "Altamaha-Ha! It's for real!" *McIntosh Life* 4, no. 3 (Fall 2017).

Spaeth, Frank. "Water Monster off the South Georgia Coast." *Fate* 51, no. 9 (September 1998).

Spitzer, Mark. *Season of the Gar: Adventures in Pursuit of America's Most Misunderstood Fish.* Fayetteville: University of Arkansas Press, 2010.


**North Fork St. Lucie River**

Butler, Douglas. "Shadowy Giant Cruising River." *Stuart News,* May 8, 1975.

"St. Lucie River 'Thing' Surfaces Once Again." *Stuart News,* May 20, 1975.

## Chapter 11 - St. Johns River Menagerie

### William Bartram's Dragon

Bartram, William. "Travels in Georgia and Florida, 1773-74: A Report to Dr. John Fothergill." *Transactions of the American Philosophical Society* n.s. 33, pt. 2 (1943).

Harper, Francis, and William Bartram. *The Travels of William Bartram.* New Haven, CT: Yale University Press, 1958.

Hoes, David. "Astor, Florida: Ground Zero for the Paranormal?" *Phantoms and Monsters*, April 26, 2012. https://www.phantomsandmonsters. com/2012/04/astor-florida-ground-zero-for.html.

Muncy, Mark, and Kari Schultz. *Eerie Florida: Chilling Tales from the Panhandle to the Keys.* Charleston, SC: The History Press, 2017.

Robinson, Robert C. *Legend Tripping: The Ultimate Adventure.* Kempton, IL: Adventures Unlimited Press, 2016.

### The Ocklawaha Serpent

"The Champion Snake." *The New York Times,* June 13, 1871.

Mitchell, C. Bradford. "Paddle-Wheel Inboard: Some of the History of Ocklawaha River Steamboats and of the Hart Line." *The American Neptune* 7, no. 2 (April 1947).

### Lake Monroe Confusions

Coleman, Loren. *Mysterious America.* London: Faber and Faber, 1983.

Firschein, Warren, and Laura Kepner. *A Brief History of Safety Harbor, Florida.* Charleston, SC: The History Press, 2013.

[Green Springs correspondent threatened]. *Pensacola Journal*, September 3, 1905.

Jenkins, Greg. *Chronicles of the Strange and Uncanny in Florida.* Sarasota, FL: Pineapple Press, 2010.

"A Liar and a Scoundrel." *Tampa Tribune*, August 31, 1905.

"Looking for a Liar." *Tampa Tribune*, September 7, 1905.

Mackal, Roy P. *Searching for Hidden Animals*. Garden City, NY: Doubleday, 1980.

Mueller, Edward A. *St. Johns River Steamboats*. Jacksonville, FL: Self-published, 1986.

[Three hundred foot serpent near Green Springs]. *Pensacola Journal*, August 24, 1905.

## The Astor Monster

Burdette, Dick. "Whatever Buck Dillard saw, it wasn't a manatee." *Orlando Sentinel*, May 9. 1982.

Burney J. Le Boeuf, and Richard M. Laws. "Elephant Seals: an introduction to the genus." in *Elephant Seals: Population Ecology, Behavior and Physiology*. Berkeley: University of California Press; 1994.

Carlson, Charlie. *Strange Florida: The Unexplained and Unusual*. New Smyrna Beach, FL: Luthers, 1997.

"Hunters Still Seek Monster." *Orlando Sentinel*, November 8. 1953.

Páez-Rosas, Diego, et al. "Southern elephant seal vagrants in Ecuador: a symptom of La Niña events?" *Marine Biodiversity Records* 11, no.13 (2018).

Powers, Ormund. "As Monsters Go, St. Johns' Was As Abominable As Any Snowman." *Orlando Sentinel*, July 19, 1995.

———. "Monster Brings Boom for Astor." *Orlando Sentinel*, October 25, 1953.

———. "River Monster Sighted Again." *Orlando Sentinel*, October 20, 1953.

———. "St. Johns Monster Sighted in Thicket by Deer Hunters." *Orlando Sentinel*, November 1, 1953.

———. "Water 'Monster' Reported in River." *Orlando Sentinel*, October 19, 1953.

"St. Johns Monster Leaves Tracks in Mud." *Orlando Sentinel*, November 21, 1953.

## Pinky and Company

Anderson, Mike. "Fisherman: Whatever It Was, It Was 20 Ft. Long." *Jacksonville Journal*, December 18, 1975.

———. "Monsters of the St. Johns an Elusive Bunch." *Jacksonville Journal*, November 23, 1976.

———. "Move Over, Nessie - Here's the St. Johns Monster." *Jacksonville Journal*, December 18, 1975.

Carlson, Charlie. *Weird Florida: Your Travel Guide to Florida's Local Legends and Best Kept Secrets*. NY: Sterling Publishing, 2005.

Coleman, Loren. "Pinky Expedition 2008." *Cryptomundo,* March 10th, 2008. https://cryptomundo.com//cryptozoo-news/pinky-08/.

———. "Pinky Expedition: Investigative Breakthrough." March 14th, 2008. https://cryptomundo.com//cryptozoo-news/pinky-carlson/.

Dalrymple, Jaclyn. "St. Johns Monster." *Tampa Tribune*, January 18, 1976.

Darby, Erasmus Foster. *A True Account of the Giant Pink Lizards of Catlick Creek Valley*. Chillicothe, OH: Ohio Valley Folktale Project, 1954.

Foley, Bill. "The River Monster of Arlington Remains a Mystery." *The Florida Times-Union* (Jacksonville), July 10, 1993.

———. "The Saga of the St. Johns River Monster Grows by the Decade." *The Florida Times-Union* (Jacksonville), October 16, 1993.

———. "Sea monster Johnny back in St. Johns." *The Florida Times-Union* (Jacksonville), December 14, 1997.

———. "Sure, it's Four Shrimp-Pink Manatees with Horns." *The Florida Times-Union* (Jacksonville), June 5, 1993.

Gunter, Gordon. "The Status of Seals in the Gulf of Mexico with A Record of Feral Otariid Seals Off the United States Gulf Coast." *Gulf Research Reports* 2, no. 3 (January 1968).

Hall, Mark A. *Natural Mysteries: Monster Lizards, English Dragons, and Other Puzzling Animals*. Minneapolis, MN: Mark A. Hall Publications and Research, 1991.

———. "Pinky, the Forgotten Dinosaur." *Wonders* 1, no. 4 (December 2002).

———. "Sobering Sights of Pink Unknowns." *Wonders* 1, no. 4 (December

2002).

Marden, William E. "Do Monsters Frolic in the St. Johns?" *The Florida Times-Union* (Jacksonville), February 19, 1989.

"'Monster' Pops Up in St. Johns." *Sentinel Star* (Orlando, FL), May 17, 1975.

"Our Monster a Sturgeon?" *The Florida Times-Union* (Jacksonville), May 17, 1975.

Paxton, Charles G. M., and Adrian J. Shine. "Consistency in Eyewitness Reports of Aquatic 'Monsters.'" *Journal of Scientific Exploration* 30, no. 1 (2016).

Puttcamp, Leland. "Listen to the Tale of Pink Lizard!" *Camerica* (*Dayton Daily News)*, November 30, 1958.

Reudiger, Steve "Pink 'Sea Monster' Lurks in River, Rattles Fishermen." *The Florida Times-Union* (Jacksonville), May 16, 1975.

"'River Monster' Reported by Florida Fishermen." *Tampa Tribune,* May 17, 1975.

"St. Johns Monster?" *St. Petersburg Times*, April 13, 1978.

Sanderson, Ivan T. "The Five Weirdest Wonders in the World." *Argosy,* November 1968.

Sass, Herbert Ravenal. "The Pink What-Is-It?" *Saturday Evening Post*, December 4, 1948.

Shuker, Karl P. N. *In Search of Prehistoric Survivors: Do Giant "Extinct" Creatures Still Exist?* London: Blandford.

Slagle, Alton. "And There was Pinky – Yecch!" *The Sunday News* (New York), June 8, 1975.

Woodley, Michael A., Darren Naish, and Hugh P. Shanahan (2008). "How many extant pinniped species remain to be described?" *Historical Biology* 20, no. 4 (December 2008).

## Chapter 12 - Cryptid Pinnipeds

Cornes, Rob, and Gary Cunningham. *The Seal Serpent.* North Devon, England: CFZ Press, 2018.

Costello, Peter. *In Search of Lake Monsters.* London: Garnstone Press, 1974.

de Muizon, Christian. *Les Vertébrés fossiles de la formation Pisco (Pérou). Première partie, Deux nouveaux Monachinae (Phocidae, Mammalia) du Pliocène de sud-Sacaco* Paris: Éditions A.D.P.E., 1981.

Mackal, Roy P. *The Monsters of Loch Ness.* Chicago: Swallow Press, 1976.

Magin, Ulrich. "St. George without a dragon: Bernard Heuvelmans and the sea serpent." *Fortean Studies* 3. (1996).

Naish, Darren. "Sea serpents, seals and coelacanths: an attempt at a holistic approach to the identity of large aquatic cryptids." *Fortean Studies* 7 (2001).

Springer, A. M., et al. "Sequential megafaunal collapse in the North Pacific Ocean: An ongoing legacy of industrial whaling?" *Proceedings of the National Academy of Sciences* 100 no 21 (October 2003).)

Walsh, Stig, and Darren Naish. "Fossil seals from late Neogene deposits in South America: a new pinniped (Carnivora, Mammalia) assemblage from Chile." *Palaeontology* 45 no.4 (July 2002).

Woodley, Michael A. *In the Wake of Bernard Heuvelmans: An Introduction to the History and Future of Sea Serpent Classification.* North Devon, UK: CFZ Press, 2008.

## The White River Monster

"220-pound Gar may be White River 'Monster.'" *Baxter Bulletin* (Mountain Home, Arkansas), January 12, 1940.

"Arkansas Has a Problem" *Pursuit* 4, no. 4 (October 1971).

"Chamber of Commerce Left Holding Bag with Unfound 'Monster' as Diver Leaves." *Arkansas Democrat* (Little Rock), July 25, 1937.

"Deep Sea Diver Has Idea; 'Monster' May Be Old Man Catfish." *Arkansas Democrat* (Little Rock), July 17, 1937.

# Bibliography

"Famed 'White River Monster' Disclosed to be a Hoax." *Baxter Bulletin* (Mountain Home, Arkansas), March 1, 1940.

Foti, Thomas. "The River's Gifts and Curses" in *The Arkansas Delta: Land of Paradox*. ed. Jeannie M. Whayne and Willard B. Gatewood. Fayetteville: University of Arkansas Press, 1993.

Harris, William. "The White River Monster of Jackson County, Arkansas: A Historical Summary of Oral and Popular Growth and Change in a Legend." *Mid-South Folklore* 5, no. 1 (Spring 1977).

Mackal, Roy P. *Searching for Hidden Animals*. Garden City, NY: Doubleday, 1980.

"Making Large Net to Snare 'Monster' Seen by Many in White River near Newport." *Weekly Citizen Democrat* (Poplar Bluff, Missouri), July 15, 1937.

Masterson, Mike. "Monster Returns to Newport After 34 Year Disappearance." *Newport Daily Independent* (Arkansas), June 18, 1971.

"Monster Sanctuary Designated." *Arkansas Gazette* (Little Rock), February 16, 1973.

"Mud Halts Search For 'Monster.'" *The Citizen-Times* (Asheville NC), July 24, 1937.

"Negro Discovers Monster Fish in White River." *Jackson County Democrat* (Newport, Arkansas), July 8, 1937.

"Newport's Monster." *Time* 30, no. 5 (August 2, 1937).

Nickell, Joe. "Arkansas's White River Monster: Very Real, but What Was It?" *Skeptical Inquirer* 42, no. 6, (November/December 2018).

"River 'Monster' is an Old Boat." *Moberly Monitor-Index* (Missouri), July 9, 1937.

"Rope Net Devised to Catch Monster." *Hope Star* (Arkansas) July 13, 1937.

"Sea Monster in River Frightens Negroes, Remains a Mystery." *Arkansas Democrat* (Little Rock), July 8, 1937.

"Something Large and Alive Lifted Boat, Two Monster Hunters Report." *Arkansas Gazette* (Little Rock), July 28, 1971.

"Three-Clawed, Fourteen Inch Tracks Found on White River Island." *Arkansas Gazette* (Little Rock), July 7, 1971.

## MS *Amerika*

Cooper, G. "The Great Sea Serpent" (letter to the editor). *Listener* 49, no. 1253 (March 5, 1953).

## St Andrew Bay Cat Serpent

Allen, Glover M. *Extinct and Vanishing Mammal of the Western Hemisphere* (Special Publication No. 11). Lancaster, PA: American Committee for International Wild Life Protection, 1942.

Burton, Maurice. *The Elusive Monster: An Analysis of the Evidence from Loch Ness.* London: Hart-Davis, 1961.

———. "The Loch Ness Saga" (in three parts). *New Scientist*, June 24, 1982, January 7, 1982, August 7, 1982.

Helm, Thomas. *Monsters of the Deep.* New York: Dodd, Mead & Co., 1962.

Shuker, Karl P. N. *The Beasts That Hide from Man: Seeking the World's Last Undiscovered Animals.* New York: Paraview Press, 2003.

## Encounter at Camp Helen

Adam, Peter J., and Gabriela G. Garcia. "New Information on the Natural History, Distribution, and Skull Size of the Extinct (?) West Indian Monk Seal Monachus Tropicalis." *Marine Mammal Science* 19, vol. 2 (2006).

Boyd, I. L., and M. P. Stanfield, "Circumstantial Evidence for the Presence of Monk Seals in the West Indies." *Oryx* 32, no. 4 (1998).

Calhoun, S. Brady. "Monster of Myth?" *The News-Herald* (Panama City, FL), October 31, 2005.

Chandler, David. "The Creature of Powell Lake Newest Bay County Mystery." *The News-Herald* (Panama City, FL), February 1, 1959.

Tohall, Patrick. "The Dobhar-Chú Tombstones of Glenade, Co. Leitrim." *The Journal of the Royal Society of Antiquaries of Ireland* 78, no. 2 (1948).

Womack, Marlene "Sea Monsters: Tales of close encounters with the wet kind" *The News-Herald* (Panama City, FL), August 25, 1985.

## The Lake Goose Waterhorse

Clark, Jerome. *Unnatural Phenomena: A Guide to the Bizarre Wonders of North America*. Santa Barbara: ABC-CLIO, 2005.

[Goose Lake monster]. *Weekly Democratic* (Austin, TX), September 15, 1881.

Newton, Michael. *Florida's Unexpected Wildlife*. Gainesville: University Press of Florida, 2007.

"Vouched for by Fishermen." *Chester Daily Times* (PA), October 25, 1881.


## Caribbean Monk Seals

Allen, J.A. "The West Indian seal (Monachus tropicalis Gray)." *Bulletin of the American Museum of Natural History* 2, no. 1 (April 1887).

Boyd, I. L., and M. P. Stanfield. "Circumstantial Evidence for the Presence of Monk Seals in the West Indies." *Oryx* 32, no. 4 (October 1998).

"Caribbean monk seal declared extinct." *Oryx* 42, no. 4 (October 2008).

Gray, John Edward. *Catalogue of the specimens of Mammalia in the collection of the British Museum*. Pt. 2, Seals. London: British Museum, 1850.

Mignucci-Giannoni, Antonio A., and Daniel K. Odell. "Tropical and Subtropical Records of Hooded Seals *(Cystophora cristata)* Dispel the Myth of the Extant Caribbean Monk Seals *(Monachus tropicalis)*." *Bulletin of Marine Science* 69, no. 1 (January 2001).

Rice, Dale. W. "Caribbean monk seal (Monachus tropicalis)." In *Seals: Proceedings of a Working Meeting of Seal Specialists On Threatened And Depleted Seals of the World, Held Under the Auspices of the Survival Service Commission of IUCN*. Morges, Switzerland: International Union for Conservation of Nature and Natural Resources, 1973.

Scheel, Dirk-Martin, et al. "Biogeography and Taxonomy of Extinct and Endangered Monk Seals Illuminated by Ancient DNA and Skull Morphology." *ZooKeys* (May 2014). doi:10.3897/zookeys.409.6244.

Solow, Andrew R. "Inferring extinction from sighting data." *Ecology* 74, no. 3 (April 1993).

Woods, Charles A., and John W. Hermanson. *An Investigation of Possible Sightings of Caribbean Monk Seals (Monachus Tropicalis), Along the North Coast of Haiti.* Washington, DC: U.S. Marine Mammal Commission, 1987.

## Chapter 13 - The Legend of Three Toes

Bothwell, Dick, "The Monster's Tracks Have Popped Up Again." *St. Petersburg Times*, October 21, 1948.

———. "Tracking the Clearwater Monster." *St. Petersburg Times*, November 14, 1948.

Brathwaite, John O. "Letters." *Strange Magazine* #18 (Summer 1997).

"City Desk" *St. Petersburg Times*, March 1, 1948.

Clarke, Arthur C. [Correspondence to Eric Frank Russell], July 27, 1954. On file with Special Collections and Archives, University of Liverpool, England.

Day, Charles William. *Five Years Residence in the West Indies.* vol 1. London: Colburn and Company, 1852.

"'Dinosaur' Got Help in Making Tracks." *New York Herald Tribune,* November 11, 1948.

Dyckman, Ruth. "'Monster' Tracks Crop Up on Courtney Campbell Parkway." *The Sun* (Clearwater, FL), June 15, 1952.

"Fact, Fiction Theories Advanced for 'Monster.'" *The Sun* (Clearwater, FL), March 7, 1948.

"Florida 'Giant Penguin' Hoax Revealed," *ISC Newsletter* 7, no. 4 (Winter 1988).

"Florida's 'Giant Penguin' Hoax Unmasked." *Pursuit* 21, no. 2 (Second Quarter, 1988).

"Florida Three Toes." *Monumental Mysteries.* Travel Channel, season 2, episode 9, August 8, 2014.

"Former Columbus Resident Tells of Sea Monster." *Columbus Dispatch,*

March 28, 1948.

Gordon, Gunter, Robert H. Williams, Charles C. Davis, and F. G. Walton Smith. "Catastrophic Mass Mortality of Marine Animals and Coincident Phytoplankton Bloom off the West Coast of Florida, November 1946 to August 1947." *Ecological Monographs* 18, no. 3 (July 1948).

Jenkins, Greg. *Chronicles of the Strange and Uncanny in Florida.* Sarasota, FL: Pineapple Press, 2010.

Joseph, E. L. *History of Trinidad.* Trinidad: Henry James Mills, 1838.

Kefauver, Evans. "'Monster's' Tracks Similar to 3-Toed Cuban Critter." *The Sun* (Clearwater), March 28, 1948.

Kirby, Jan. "Clearwater Can Relax: Monster Is Unmasked." *St. Petersburg Times,* June 11, 1988.

Klinkenberg, Jeff. "Man, not Beast." *St. Petersburg Times,* June 26, 2006.

"'Monster' Has Now Moved to Indian Rocks." *The Sun* (Clearwater), March 21, 1948.

"Mysterious Critter Leaves Tracks on Beach; Name It and You Can Have It." *The Sun* (Clearwater), February 29, 1948.

Newton, Michael. *Florida's Unexpected Wildlife.* Gainesville: University Press of Florida, 2007.

O'Reilly, John. "3-Toed Tracks like a Dinosaur's Keep Suwannee Land Stirred Up." *New York Herald Tribune,* November 10, 1948.

Pittman, Craig. "Feet Leave Lasting Impression." *St. Petersburg Times,* January 5, 2014.

Rickard, Bob. "Florida's Penguin Panic," *Fortean Times* 66 (December 1992-January 1993).

Rigsby, G. G. "It's Still a Monster of a Tale." *St. Petersburg Times,* October 30, 1998.

Robinson, Robert. "The Clearwater Monster Tracks." *Legend Tripping,* November 17, 2015. https://legendtrippersofamerica.blogspot.com/2015/11/the-clearwater-monster-tracks.html.

Sanderson, Ivan T. *Investigating the Unexplained: A Compendium of Disquieting Mysteries of the Natural World.* Englewood Cliffs, N.J.: Prentice-Hall, 1972.

——. "That Forgotten Monster - Old Three-Toes." *Fate* 20, no 12 and 21, no. 1 (December 1967, January 1968). Reprinted in *More "Things."* NY: Pyramid Books, 1969.

——. "The Tracks." *True,* June 1951.

——. "There Could be Dinosaurs." *Saturday Evening Post*, January 3, 1948.

"Sanderson Reports." *Doubt* 24 (April 1949).

"'Sea Monster' Runs on Beach." *Miami Herald,* February 29, 1948.

"Science: Captive Red Tide." *Time,* April 9, 1956.

Shurtleff, Ted. "That Monster's Here Again; 4 Fliers Say They Saw It." *St. Petersburg Times*, July 26, 1948.

"That Monster is Back Again." *St. Petersburg Times*, March 21, 1948.

"This Time the Monster is Seen Swimming." *The Sun* (Clearwater), July 26, 1948.

"Thousands Flock to 'Footprints.'" *St. Petersburg Times*, March 23, 1948.

Trumbull, Stephen. "'Dinosaur Tracks' Have a Press Agent Smell." *Miami Herald,* November 24, 1948.

## Chapter 14 - Early Encounters

Juan, Jorge, and Antonio de Ulloa. *Relacion Historica del Viage a la America Meridional*, vol. 1. Madrid: Antonio Marín, 1748.

López de Gómara, Francisco. *La Historia General de las Indias.* Anvers: Iuan Steelsio, 1554.

Philoponus, Honorius (Caspar Plautius). *Nova typis transacta navigatio.* [Linz]: n.p., 1621.

## Giant Turtles

Armas, Juan Ignacio de. *La zoología de Colón y de los primeros exploradores de América.* Habana: Establecimiento tipógrafico, 1888.

"First Sea Serpent Lies on Way to Cuba." *Evening World* (New York), April

27, 1921.

France, Robert. "Historicity of Sea Turtles Misidentified as Sea Monsters: A Case for the Early Entanglement of Marine Chelonians in Pre-plastic Fishing Nets and Maritime Debris." *Coriolis: the Interdisciplinary Journal of Maritime Studies* 6, no. 2 (2016).

Las Casas, Bartolomé de, *Historia de las Indias*, vol. 2. Madrid: Imprenta de Miguel Ginesta, 1875.

Magin, Ulrich. "In the Wake of Columbus' Sea Serpent: The Giant Turtle of the Gulf Stream." *Pursuit* 20, no. 2 (whole number 78, second quarter 1987).

Márquez, Gabriel García. *Relato de un náufrago*. Buenos Aires: Editorial Sudamericana, 1970.

Morison, Samuel Eliot. *Admiral of the Ocean Sea,* vol 2. Boston: Little, Brown and Company, 1942.

## De Orbe Novo

d'Anghiera, Pietro Martire. *De Orbe Nouo Petri Martyris ab Angleria Mediolanensis protonotarij cesaris senatoris decades*. Compluti: Apud Michaelé d' Eguia, 1530.

de Bry, Theodor. *Americæ pars VIII*. Francfort: Opera & sumptibus Theodorici de Bry P.M. relictae Viduae & filiorum, 1599.

Raynal, Michel, and Michel Dethier. "Le 'Monstre de Floride' de 1896: Cétacé ou Poulpe Colossal?" *Bulletin de la Société Neuchâteloise des Sciences Naturelles* 114 (1991).

## The Marrajo of Las Casas

Castro, José I. "Historical Knowledge of Sharks: Ancient Science, Earliest American Encounters, and American Science, Fisheries, and Utilization." *Marine Fisheries Review* 75, no. 4 (2013).

———. "On the origins of the Spanish word 'tiburón,' and the English word 'shark.'" *Environmental Biology of Fishes* 65, no. 3 (November 2002).

Jones, Tom. "The *Xoc*, the *Sharke*, and the Sea Dogs: An Historical Encounter." in *Fifth Palenque Round Table, 1983*. ed. Virginia M. Fields. San Francisco: Pre-Columbian Art Research Institute, 1985.

Las Casas, Bartolomé de, *Historia de las Indias*, vol. 3. Madrid: Imprenta de Miguel Ginesta, 1875.


**Maps and Monsters**

le Moyne, Jacques, Theodor de Bry, and Carolus Clusius. *Brevis narratio eorum quae in Florida Americae provincia Gallis acciderunt, secunda in illam navigatione, duce Renato de Laudonniere classis praefecto anno MDLXIIII. quae est secunda pars Americae.* Francoforti ad Moenum: typis Joannis Wecheli, 1591.

Magnus, Olaus. Historia De Gentibus Septentrionalibus. Farnborough, England: Gregg, 1971 (Facsimile reprint of 1st ed., Romae: J.M. de Viottis, 1555.).

Van Duzer, Chet. *Sea Monsters on Medieval and Renaissance Maps.* London: The British Library, 2014.


**Chapter 15 - Folklore and Fakelore**


Heuvelmans, Bernard. "What is Cryptozoology?" *Cryptozoology* 1 (Winter 1982).

Krueger, Charles. "Ocean Bathers Startled by Black, Shiny Object." *Miami Daily News*, July 28, 1949.


**Pablo Beach Problem**

"Alarming Experience of Fair Bathers Who Are Attacked by an Octopus." *The Illustrated Police News*, October 17, 1896.

"Carried to Sea by a Turtle." *The Evening News* (New York), July 21, 1891.

Jenkins, Greg. *Chronicles of the Strange and Uncanny in Florida.* Sarasota, FL: Pineapple Press, 2010.

## The Soldier Key Skull

"Calls Monster Whale, Not Fish." *Miami Herald,* February 13, 1921.

"Find Huge Sea Monster." *New York Times,* February 13, 1921.

"Fragment of Sea Monster, Cast Ashore, Brought Here." *Miami Herald,* February 12, 1921.

"Skull of Monster 'Sea Serpent' Brought Here Puzzles Scientists." *Miami Herald,* February 24, 1921.


## Sanibel Island

Adler, Elizabeth Thompson. *Wakulla Springs: Its History, Legend, Birds and Wildlife.* [Wakulla Springs, FL]: Wakulla Silver Springs Co, 1977.

Carlin, Patricia. "Most Credible Sea Monster or Giant Oarfish Proof ever caught on video!! Must watch." Filmed Oct 6, 2013. YouTube video, 00:54. Posted October 8, 2013. https://youtu.be/c1G76AyW1ag.

"Sanibel Monster?" *Island Reporter* [Sanibel, FL], July 19, 1974.


## Normandy Nessie

Challenger, Rod. "Are Sea Monsters Invading Canals?" *Tampa Tribune,* November 17, 2009.

Estrada, Sheila Mullane. "'Big beast' Reported in Canal." *St. Petersburg Times,* November 15, 2009.

"Florida 'Nessie' Lurking in Canals?" *St. Petersburg Times,* November 14, 2009.

Rudie, Preston. "Mysterious sea creature spotted off Sand Key." WSTP-TV, September 1, 2010. https://www.wtsp.com/article/home/mysterious-sea-creature-spotted-off-sand-key/67-390395457.

Sittloh, Russell. [Normandy Nessie footage]. YouTube Videos, 1-5. November 8, 10, 14, 2009. https://www.youtube.com/user/Apollo6818.


## Great Horned Reptiles

Bardet, Nathalie. "Evolution et Extinction des Reptiles Marins au cours du Mésozoïque." *Palæoverbrata* 24, no. 3-4 (October 1995).

Cushing, Frank Hamilton. "Exploration of Ancient Key-Dwellers' Remains on the Gulf Coast of Florida." *Proceedings of the American Philosophical Society* 35, no. 153 (December 1896).

Howard, James H., in collaboration with Willie Lena. *Oklahoma Seminoles: Medicines, Magic, and Religion.* Norman: University of Oklahoma Press, 1984.

Lankford, George E. "The Great Serpent in Eastern North America." in *Ancient Objects and Sacred Realms.* ed. F. Kent Reilly III and James F. Garber. Austin: University of Texas Press. 2010.

Lenik, Edward J. "Mythic Creatures: Serpents, Dragons, and Sea Monsters in Northeastern Rock Art." *Archaeology of Eastern North America* 38 (2010).

Naish, Darren. "Sea Serpents, Seals, and Coelacanths." *Fortean Studies* 7 (2001).

Smith, Theresa S. *The Island of the Anishnaabeg: Thunderers and Water Monsters in the Traditional Ojibwe Life-World.* Moscow, Idaho: University of Idaho Press, 1995.

## Chapter 16 -Mermaids, Mermen, and Manatees

Beck, Horace. *Folklore and the Sea.* Mystic CT: Mystic Seaport Museum, 1973.

Carrington, Richard. *Mermaids and Mastodons: A Book of Natural and Unnatural History.* NY: Rinehart & Company, 1957.

Clark, Jerome. *Unexplained! Strange Sightings, Incredible Occurrences & Puzzling Physical Phenomena* (third ed.). Detroit: Visible Ink, 2012.

Drewal, Henry John. "Interpretation, Invention, and Re-Presentation in the Worship of Mami Wata." *Journal of Folklore Research* 25, no. 1/2 (January-August, 1988).

### Columbus' Sirenas

Morison, Samuel Eliot. *Admiral of the Ocean Sea: A Life of Christopher Columbus*, v.1. Boston: Little, Brown & Co, 1942.

Smith, Virginia. "Sea Cow 'Sirens' Fuel Mermaid Mythology." *The Daytona Beach News-Journal*, December 25, 2005.


**The Hipupiára**

Cardim, Fernão, *Tratados da terra e gente do Brasil*. Rio de Janeiro: J. Leite & Cia, 1925.

Gândavo, Pêro de Magalhães de. *História da Província Santa Cruz*. Ed. Clara C. Souza Santos, and Ricardo M. Valle. São Paulo: Hedra, 2008.

Léry, Jean de. *Histoire d'un Voyage fait en la Terre du Brésil: dite Amérique*. A La Rochelle: Antoine Chuppin, 1578.

Whitehead, P. J. P. "Registros antigos da presença do Peixe-Boi do Caribe (Trichechus manatus) no Brasil." *Acta Amazonica* 8, no. 3 (July/September 1978).


**John Smith & the Green-Haired Mermaid**

de Bry, Theodore. *Dreyzehender Theil Americae*. Franckfurt: Caspar Rötel, 1628.

Dumas, Alexandre. "Les Mariages du Père Olifus (chapter one)." *Le Constitutionnel,* July 10, 1849.

———. "Nuptials of Father Polypus (chapter one)." *Gazette of the Union*, September 29, 1849.

Landrin, Armand. *Les Monstres Marins*. Paris: Librairie de l. Hachette et Cie, 1877.

Scribner, Vaughn. "Fabricating History: The Curious Case of John Smith, a Green-Haired Mermaid, and Alexandre Dumas." *The Junto,* June 15, 2015. https://earlyamericanists.com/2015/06/16/guest-post-from-vaughn-scribner-fabricating-history-the-curious-case-of-john-smith-a-green-haired-mermaid-and-alexandre-dumas/.

———. "Fabricating History Part Two: The Curious Case Continues." *The Junto,* July 2, 2015. https://earlyamericanists.com/2015/07/02/fabricating-history-part-two-the-curious-case-continues/.

Smith, John. *The Complete Works of John Smith*. 3 vols. ed. Thad W. Tate and Philip L. Barbour. Chapel Hill: University of North Carolina Press, 1986.

Stengel, George. *De Monstris et monstrosis quam mirabilis, bonus et justus, in mundo administrando, sit Deus, monstrantibus*. Ingolstadii: Gregory Hænlin, 1647.

Whitbourne, Richard. *A Discourse and Discovery of New-Found-Land*. London: Felix Kingston, 1620.


**Benjamin Franklin's Contribution**

"Philadelphia, April 29." *The Pennsylvania Gazette*, April 22-29, 1736.


**The van Batenburg Mermaids**

Chisholm, C. *An Essay on the Malignant Pestilential Fever, Introduced into the West Indian Islands from Boullam, on the Coast of Guinea, as it appeared in 1793, 1794, 1795, and 1796,* vol. 2, rev. ed. London: J. Mawman, 1801.

Stedman, J. G. *Narrative of a five years expedition against the revolted Negroes of Surinam,* vol. 2. London: J. Johnson, 1796.


**The schooner *Addie Schaeffer***

"A Mermaid Captured." *Evening Metropolis* (Jacksonville, FL), April 19, 1890.


**Mr. Carruther's Mermaid**

"He Did See a Mermaid." *Brooklyn Daily Eagle*, October 7, 1894.


**Captain Lowe's Mermaid**

Cappick, Marie. "Key West Resident Sees a Mermaid." *Paths* 1, no. 5 (July 1934).

Cappick, Marie. "Three-Year Search for Man Who Saw Mermaids Ends on Porch of Key West Dwelling." *Havana-American News,* July 20, 1930.

## An Atlantean Mermaid

Floyd, E. Randall. *Great Southern Mysteries: Bigfoot, Water Witches, the Carolina Bays, the Biloxi Mermaid, and Other Enduring Mysteries of the Deep South.* AR: August House, 1989.

## Water Apes

Coleman, Loren, and Patrick Huyghe. *The Field Guide to Bigfoot and Other Mystery Primates.* San Antonio, TX: Anomalist Books, 2006.

Hall, Mark A., and Loren Coleman. *Merbeings: The True Story of Mermaids, Mermen, and Earth's Lizard People.* Portland, ME: International Cryptozoology Museum, 2020.

Heuvelmans, Bernard. "Annotated Checklist of Apparently Unknown Animals with which Cryptozoology is Concerned." *Cryptozoology* 5 (1985).

Köhler, Meike, and Salvador Moyà-Solà. "Ape-like or hominid-like? The positional behavior of *Oreopithecus bambolii* reconsidered." *Proceedings of the National Academy of Sciences* 94 (Oct 1997).

## Epilogue

Bancroft, Hubert Howe. *The Native Races of the Pacific States of North America.* vol 3. New York: D. Appleton, 1875.

Bowles, Richard. "Most Seamen Sight a Sea Monster Now and Then." *Gainesville Sun* (Florida), February 20, 1987.

Brünnich, M. Thr. *Zoologiæ Fundamenta Prælectionibus Academicis Accommodata.* Copenhagen: Frederik Christian Pelt, 1772.

"'Fish Stories' are in Vogue at Indian Rocks." *Tampa Tribune*, March 17, 1915.

"Fish Stories Originating in Punta Gorda are True – Jordan." *Tampa Tribune*, April 10, 1915.

Gould, Rupert T. *The Loch Ness Monster and Others.* Secaucus, NJ: Citadel Press, 1976.

"The 'Great Sea-Serpent' Found at Last." *Harper's Weekly*, March 3, 1860.

Helm, Thomas. *Monsters of the Deep.* New York: Dodd, Mead & Co., 1962.

Monk, Jerry. [Letter to the editor]. *Fortean Times* (July 2004).

"More on the 'Great Gymnetrus' or Sea Serpent." *The New York Times*, February 21, 1860.

Orihara, Yoshiaki, et al. "Is Japanese Folklore Concerning Deep-Sea Fish Appearance a Real Precursor of Earthquakes?" *Bulletin of the Seismological Society of America* 109, no. 4 (August 2019).

Radford, Benjamin, and Joe Nickell. *Lake Monster Mysteries - Investigating the World's Most Elusive Creatures.* Lexington: University Press of Kentucky, 2006.

Roberts, Tyson R. *Systematics, biology, and distribution of the species of the oceanic Oarfish genus Regalecus.* Paris: Publications Scientifiques du Muséum, 2012.

Snow, Ray W., et al. "Introduced Populations of *Boa Constrictor* (Boidae) and *Python molurus bivittatus* (Pythonidae) in Southern Florida" in *Biology of the Boas and Pythons.* ed. Robert W. Henderson and Robert Powell. Eagle Mountain, UT: Eagle Mountain Publishing, 2007.

Spitzer, Mark. *Season of the Gar: Adventures in Pursuit of America's Most Misunderstood Fish.* Fayetteville: University of Arkansas Press, 2010.

"Sunday Excursionists Startled by Sea-Beast." *The Weekly Tribune* (Tampa), August 29, 1912.

Winter, Nevin O. *Florida: The Land of Enchantment.* Boston, Mass: The Page Company, 1918.

Zarzynski, Joseph W. *Champ, Beyond the Legend.* Port Henry, NY: Bannister Publications, 1984.

# INDEX

Printed in the USA
CPSIA information can be obtained
at www.ICGtesting.com
JSHW010443180324
59284JS00004B/7

9 781949 501117